THE SCIENCE OF
WITCHCRAFT

THE TRUTH BEHIND SABRINA, MALEFICENT, GLINDA, AND MORE OF YOUR FAVORITE FICTIONAL WITCHES

MEG HAFDAHL & KELLY FLORENCE

AUTHORS OF *THE SCIENCE OF WOMEN IN HORROR*

Skyhorse Publishing

Skyhorse Publishing books may be purchased in bulk at special discounts for sales promotion, corporate gifts, fund-raising, or educational purposes. Special editions can also be created to specifications. For details, contact the Special Sales Department, Skyhorse Publishing, 307 West 36th Street, 11th Floor, New York, NY 10018 or info@skyhorsepublishing.com.

Skyhorse® and Skyhorse Publishing® are registered trademarks of Skyhorse Publishing, Inc.®, a Delaware corporation.

Visit our website at www.skyhorsepublishing.com.

10 9 8 7 6 5 4 3 2 1

Library of Congress Cataloging-in-Publication Data is available on file.

Cover design by David Ter-Avanesyan
Cover images by Shutterstock

Print ISBN: 978-1-5107-6718-8
Ebook ISBN: 978-1-5107-6719-5

Printed in the United States of America

We dedicate this book to the witches of the past, present, and future who used their powers for good.

CONTENTS

INTRODUCTION

Double, double toil and trouble;
Fire burn and caldron bubble.
Fillet of a fenny snake,
In the caldron boil and bake;
Eye of newt and toe of frog,
Wool of bat and tongue of dog,
Adder's fork and blind-worm's sting,
Lizard's leg and howlet's wing,
For a charm of powerful trouble,
Like a hell-broth boil and bubble.
—William Shakespeare

Shakespeare's words are timeless. While they were written on parchment hundreds of years ago, they still resonate today. This "Song of the Witches" from *Macbeth* is no different, casting an indelible spell on our understanding of the secret life of witches. The classic media depiction of witches dressed in black, circling a cauldron bubbling with animal parts has endured in everything from goofy cartoons to terrifying horror films. Over time, the story of the witch has come alive in a million different forms: good, bad, spiritual, demonic, cannibalistic, healing, young, and old. In this book, we aim to pull back the veil of stereotypes and reveal the true history, legends, and science that inhabit the mystical world of witches.

Now, kiddies, follow along if you dare! (Meg and Kelly cackle and disappear in a plume of smoke.)

SECTION ONE
THE ORIGINS

CHAPTER ONE
THE WIZARD OF OZ

Say the word "witch" and chances are you'll conjure up an image of a woman wearing a pointy black hat (cackling maniacally, of course). The word itself has taken on many iterations over the centuries. The first mention of witches in the Hebrew Bible appears in the first book of Samuel (28:3–25), which refers to a story about a woman performing necromancy and magic. Writers in the fifteenth century used the term *maleficus*, meaning a person who "performed harmful acts of sorcery against others."[1] However, witches became known as "women whose embodiment of femininity in some way transgresses society's accepted boundaries; they are too old, too powerful, too sexually aggressive, too vain, too undesirable."[2]

> The earliest known written magical incantations come from ancient Mesopotamia (modern Iraq), dated to between the 5th and 4th centuries BCE.[3]

Witches have come a long way on screen from *The Maid of Salem* (1937) to *Wandavision* (2021–). Perhaps the most famous portrayal is the Wicked Witch from *The Wizard of Oz* (1939). "I'll get you, my pretty, and your little dog too!" This iconic line, spoken by the Wicked Witch of the West (Margaret Hamilton) in the film, epitomizes the witch villain. She has a horrific green face, cackles wildly, and is not only willing to hurt a young girl but her lovable pooch too! How could we possibly empathize with this terrible, monstrous woman? We can and do through the journey that is the musical *Wicked* (2003). Created by Stephen Schwartz and Winnie Holzman, *Wicked* is the story of Elphaba and Galinda (later Glinda), witches in the Land of Oz.

The journey from page to stage took many iterations over the course of one hundred years. The original book, *The Wonderful Wizard of Oz* by L. Frank Baum, is considered the first American fairy tale. Baum, who wrote fourteen novels in the Oz series, was inspired by places and people from his childhood, including a recurring nightmare in which he was chased across a field by a scarecrow.[4] The book was considered a success but came under fire decades later by fundamentalist Christians for its inclusion of, you guessed it, witches. In 1986, parents in the Hawkins County, Tennessee, school system tried to get the book banned in their local school due to its depiction of witches as benevolent beings.[5] This wasn't the last time parents

During the medieval times of Britain, scarecrows originated as young children who would throw stones when birds landed in fields. When the Great Plague killed nearly half of Britain's population in 1348, they switched to scarecrows being made out of stuffed sacks of straw.[6]

protested a movie or book series out of fear that their children would turn to witchcraft. A vocal group of Christians based in Colorado Springs, Colorado, fought against the Harry Potter books (1997–2007) being in school[7] and a group in Pennsylvania even staged burnings of the series.[8] What is the psychology behind this? Parents who fear the inclusion of witchcraft in their children's stories often believe that by reading these stories their kids will turn to the occult and begin practicing spells themselves. Experts say that's not how reading works, though. This was demonstrated in a study by psychologists Amie Senland and Elizabeth Vozzola:

In their study comparing the perceptions of fundamentalist and liberal Christian readers of *Harry Potter*, Senland and Vozzola reveal that different reading responses are possible in even relatively homogeneous groups. On the one hand, despite adults' fears to the contrary, few children in either group believed that the magic practiced in *Harry Potter* could be replicated in real life. On the other, the children disagreed about a number of things, including whether or not Dumbledore's bending of the rules for Harry made Dumbledore harder to respect. Senland and Vozzola's study joins a body of scholarship that indicates that children perform complex negotiations as they read. Children's reading experiences are informed by both their unique personal histories and their cultural contexts. In other words, there's no "normal" way to read *Harry Potter*—or any other book, for that matter.[9]

Scarecrows began in the fields of ancient Greece as wooden statues carved to represent Priapus, a Greek god of fertility.[10]

The Wizard of Oz came out in 1939 and is considered one of the American Film Institute's (AFI's) one hundred greatest American movies of all time.[11] It was nominated for six Academy Awards and went on to spur two film sequels: *Journey Back to Oz* (1972) and *Return to Oz* (1985). I (Kelly) spent many days watching the original movie on my well-worn VHS tape hoping for a tornado to sweep me away to the Land of Oz. Thankfully, no tornado ever came but my imagination soared while watching this story. It was often pointed out to me that the film's star,

Judy Garland, grew up not far from where I did. We often visited her home in Grand Rapids, Minnesota, as it has been turned into a museum. Many kids I knew liked the movie but were *terrified* of the wicked witch. How did Margaret Hamilton come to portray her character?

Hamilton was an actress who gained credits in local theatre and several films before she auditioned to be the Wicked Witch of the West. Some people in the industry told her to get a nose job if she ever expected to be cast. She didn't, though, and got over what other people thought of her looks. That confidence no doubt carried over into her audition and she got the part. Hamilton never missed a chance to help children see the witch's human side. Having been a kindergarten teacher, she understood children and knew it was important to let them know the witch wasn't real. She spoke to Fred Rogers on *Mister Rogers' Neighborhood* (1968–2001) and said about her character, "she's very unhappy because she never gets what she wants, Mr. Rogers. Most of us get something . . . but as far as we know, that witch has never got what she wanted."[12]

What medical conditions could explain the witch's green skin? A condition called hypochromic anemia was historically known as green sickness for the distinct skin color sometimes present in patients. Other symptoms from this type of anemia include lack of energy, shortness of breath, headaches, and low appetite. It's caused by a lack of vitamin B6, certain infections, or diseases. Other conditions that have been known to cause green skin include multiple organ failure, contact with copper, or the use of certain drugs.[13]

Dorothy stops at Professor Marvel's (Frank Morgan's) wagon after she runs away in the beginning of *The Wizard of Oz*. He uses a crystal ball to tell her about her life and, later, the Wicked Witch uses one to spy on Dorothy. What are crystal balls used for in witchcraft? Witch balls in folklore were used for scrying, the act of fortune-telling or divination. The ball may give guidance, inspiration, or clarity to the user, allowing them to see faraway places or predict future events. Scrying could also involve an object other than a crystal ball, including other reflective surfaces like water or a mirror, or objects in nature like smoke or the movement of animals. The scryer will stare at the chosen object until they are in a trance-like state, allowing them to see visions in the object or in their head. Nostradamus was said to scry when he came up with his predictions for the future.[14] Can science explain the effects of this phenomenon? Experts say it may be a combination of delusion or wishful thinking. One study of mirror gazing concluded it "might be caused by low-level fluctuations in the stability of edges, shading, and outlines affecting the perceived definition of the face, which gets overinterpreted as 'someone else' by the face recognition system."[15] The participants of the study reported feeling strong emotions while mirror gazing, though, which proves that the mind and our perceptions are quite powerful. This could explain why we "see" faces in the pattern of a kitchen countertop, eyes in a tree, or a predominant shape in a cloud.

After Dorothy visits Professor Marvel, she is swept up in the tornado that brings her to Oz. Can tornadoes transport people, objects, or even houses to new places? Paper debris caught in a cyclone has been known to travel up to two hundred miles away while some heavier objects have traveled around fifty miles.[16] In 2017, a woman in Texas climbed inside her bathtub to take shelter from an impending storm. The tornado ripped the bathtub, with her in it, from the home and deposited it in the nearby woods. She walked away relatively unharmed with only some cuts and bruises![17] In 2020, a tornado in Georgia tore an entire house off its foundation, lifted it in the air, and dropped it on a highway fifty yards away.[18] Thankfully, no one was inside. The United States averages about one thousand tornadoes per year with ninety-six of those being in Dorothy's home state of Kansas.

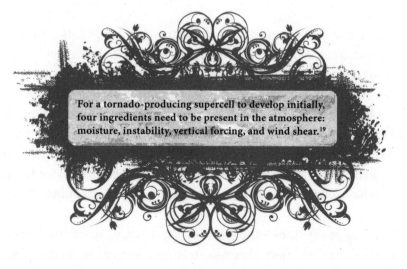

For a tornado-producing supercell to develop initially, four ingredients need to be present in the atmosphere: moisture, instability, vertical forcing, and wind shear.[19]

After Dorothy meets up with her newfound friends and is about to enter the Emerald City, The Wicked Witch uses poppies to put them to sleep. From the book:

> Now it is well known that when there are many of these flowers together their odor is so powerful that anyone who breathes it falls asleep, and if the sleeper is not carried away from the scent of the flowers, he sleeps on and on forever. But Dorothy did not know this, nor could she get away from the bright red flowers that were everywhere about; so presently her eyes grew heavy and she felt she must sit down to rest and to sleep ... "If we leave her here she will die," said the Lion. "The smell of the flowers is killing us all."[20]

What is the science behind this? According to Joe Schwarcz, PhD, it's not scientifically possible that this could have worked:

> The Latin botanical name of the flower, Papaver somniferum, translates as 'sleep-bringing poppy.' But smelling poppies is not enough to bring on sleep, as the active components are not volatile. Ingestion or injection of "opiates" is required. Opiates are biologically active compounds extracted from opium, the dried latex that exudes from an incision made in the seed pods of the plant before these blossom into flowers.[21]

Opium use peaked in the United States just around the time *The Wonderful Wizard of Oz* was written so it was undoubtedly on the mind of the author and the public.

The book *Wicked: The Life and Times of the Wicked Witch of the West* (1995) by Gregory Maguire was adapted to a stage musical in 2003. Told from the perspective of the witches, numerous topics are explored, including discrimination, animal cruelty, and the nuances behind good versus evil, friendship, and love. Idina Menzel, who originated the role of Elphaba on Broadway, believed the play had important messages:

> The two main ones—the story of the friendship of these two women and how they, through striving for truth in themselves,

they really give a wonderful gift to each other and change each other forever. I think that that's important. I think that the idea that when someone's different from us, we tend to be threatened by them, and that we have to strive to look deeper than the surface. I think that's the other most important message.[22]

The idea of good versus bad regarding magic and witches comes up in numerous stories throughout history and various religions.

Female witchcraft, for instance, surprisingly is mentioned only rarely in early Christian narrative texts, though male sorcerers are quite common. Women in such stories usually are victims of witchcraft, not perpetrators. A popular and naïve belief in demons is still often seen as part of a common antique "weltbild," [a theoretical understanding of the world and how it operates] shared and indeed intensified by Christians. Once again, things become much more complicated when the full range of sources is allowed to speak. Jewish views differ substantially from Christian ones, and there are even Christian as well as Jewish sources that use terms like "magician" (magos) in surprisingly positive contexts. Ideas of magic and witchcraft in Christian history have changed much, and not only for theological reasons. The New Testament as a collection of foundational Christian writings only rarely deals in magic as such, though Jesus himself was accused by his opponents as someone in league with Beelzebul.[23]

Good versus bad can also be viewed as white versus dark magic. White magic is often associated with performing beneficial acts for others and black magic typically relates to curses and nefarious intentions.

Although the Wicked Witch of the West died unintentionally near the end of *The Wizard of Oz*, she made a lasting impression on generations of people and will no doubt continue to do so for years to come.

CHAPTER TWO
SLEEPY HOLLOW

The Legend of Sleepy Hollow by Washington Irving was first published in 1819 in his book of short stories *The Sketch Book*. He was inspired by the legends of headless horsemen he heard of while in Europe, including the Dullahan, an Irish myth of a man carrying around his own moldy, wide-grinning head on a black horse. The town of Sleepy Hollow in the story is said to have been bewitched. Its inhabitants experience all sorts of supernatural and mysterious occurrences, including visits from the Headless Horseman.

Is it possible for humans or animals to run around headless like the Headless Horseman in this story? It turns out, there is a bizarre truth to the old adage "running around like a chicken with your head cut off." Chickens do indeed run around after decapitation. In one miraculous instance, a chicken ran around for eighteen months. How is this possible? In 1945, a chicken named Mike was seemingly decapitated but was found alive the next day. According to *Modern Farmer*'s Rebecca Katzman, "Chicken's brains are arranged at such an angle that the most basic parts of the brain, the cerebellum and the brain stem, can remain nestled in the neck even if most of the head is gone."[1] The owners fed Mike through an eye dropper, but he eventually died in a hotel room while on tour. What a rock star! Humans cannot live without their heads but what about the other way around? Does a decapitated head remain conscious for a period of time? A study from 2011 found the following:

There is an annual "Mike the Headless Chicken Festival" typically held the first weekend in June in Mike's hometown of Fruita, Colorado. The festival includes vendors, a 5K run, and a disc golf tournament.[2]

Mice, whose scalps were hooked on electroencephalography machines, preserved electrical activity whose frequencies indicated conscious activity for nearly four seconds after the mice were decapitated. Other studies of small mammals have found up to twenty-nine seconds. If this is also valid for humans, then it would provide enough time for a most gruesome experience.[3]

No, thank you!

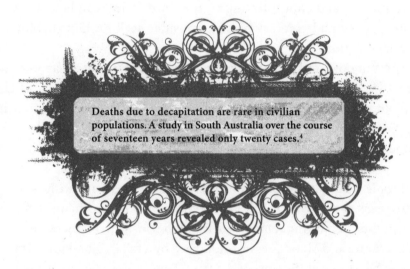

Deaths due to decapitation are rare in civilian populations. A study in South Australia over the course of seventeen years revealed only twenty cases.[4]

There are numerous iterations of *The Legend of Sleepy Hollow*, including the 1949 Disney film entitled *The Adventures of Ichabod and Mr. Toad*. Tim Burton took on directing the 1999 version (*Sleepy Hollow*) and cast Johnny Depp as Ichabod Crane, a nervous detective in this adaptation. As Ichabod enters a party, jack-o'-lanterns are seen everywhere. What is the history behind these common Halloween mainstays? It began with the Irish legend of Stingy Jack who tricked the Devil one too many times and ended up walking the earth with only coal placed inside a carved-out turnip lighting his way. He became known as "Jack of the Lantern" and people would carve scary faces into turnips or potatoes to keep him and other evil spirits away. When the tradition was brought to the United States, it was found that pumpkins were the perfect vehicle for these scary faces.

Ichabod begins to shake when he hears the tale of the Headless Horseman. Why do our bodies do this? Over forty million Americans experience anxiety and when we feel stress, our bodies often go into fight or flight mode. This means our adrenaline kicks in, our heart rate increases, and our breathing quickens. When our hands or other parts of our bodies start to shake, this is considered a psychogenic tremor. How can we treat them? First, progressive muscle relaxation is recommended. Contracting and releasing various muscle groups in your body can help it calm down and stop any shaking. Second, try to slow your breathing and practice mindfulness. Being aware of your surroundings and concentrating on the word "calm" can help. Finally, if the anxiety and tremors persist, consider seeking talk therapy or even medication.

Speaking of panic, Ichabod passes out from fear more than once in the movie. Does this happen in real life? Absolutely! It's that darn fight or flight response again that's sending more blood to your legs to help you flee but pumps less blood to your head. This same response may happen at the sight of blood. According to Dr. Sue Corcoran:

> People who faint at the sight of blood may also faint with pain triggers such as an injury or vaccination. The mechanism for this type of fainting remains elusive. It begins with an increase in heart rate and blood pressure, followed by a fall in blood pressure and heart rate at the time of the faint. People often feel nauseated before they faint (thought to be due to the back of the brain not getting enough blood supply) and may vomit.[5]

The tree in *Sleepy Hollow* appears to be weeping blood. We find out later it's due to the stash of decapitated heads the Horseman is keeping there! What could really cause a tree to weep in nature? There is a tree in South Africa that appears to "bleed" when cut. The appropriately named bloodwood tree naturally has dark, red sap. Locals believe the tree to have magical, healing powers and use its sap to treat health issues such as ringworm. Many trees in the world produce sap, some edible and some that show signs of disease, damage, or pests. It's important to do your research before ingesting any sap from a tree that you're unfamiliar with.

Wild arrowroot flowers are used to heal a cut in the film. What other natural cures are there? We had the opportunity to interview Indigenous ethnobotanist Linda Black Elk about her experience and training using natural remedies.

Meg: **"Tell us how you learned about indigenous healing."**
Linda Black Elk: "I currently work as the food sovereignty coordinator at United Tribes Technical College, the tribal college in Bismarck, North Dakota. Before that, for twenty years I taught at Sitting Bull College, the tribal college on the Standing Rock Reservation. My two eldest sons are enrolled at Standing Rock and my husband and I and our son are enrolled on the Cheyenne River Reservation, which borders Standing Rock to the south. I've lived around Lakotas (the people of Standing Rock) for a long time and learned so much from them. I've talked to so many elders and other knowledge holders on Standing Rock and Shine River and in Lakota Country. It was always a huge passion of mine. I was always interested in plants for food and medicine. My grandmother would take me on plant walks all the time. She knew so much about plants. I've been foraging and eating plants and using plants for medicine my whole life. When I started college (no one in my family had ever gone to college before), I didn't know what to major in. Plants and people seemed to be the logical thing, so I got into anthropology, botany, and literature as well. I even joke around sometimes that I get along much better with plants than I do people!"
Meg: "I feel that way about pugs!"
Linda Black Elk: "I've always had very good instincts about plants. I personally consider plants as relatives, not just these things that grow out there. I have a very close relationship with plants, and I consider them to be very communicative. I follow the protocols that I've always been taught to follow when I'm going out to talk to plants and harvest and things like that."

Kelly: **"You say that you speak to plants. Tell us more about that."**
Linda Black Elk: "Sure! So many traditional protocols are based in science and on the need to protect the plant or to harvest it at a particular time. Maybe when it's at its maximum potential. As someone who's also trained as a Western scientist (my masters is in science education), those ideas are really important to me. I do talk to plants, and I believe that they speak

back to me. People consider this to be a very mystical sort of thing, but I find it to be very practical. For example, if I'm harvesting a medicine and I find that where I've normally collected tons of it, it's not available this year, then I'm not supposed to harvest there. That's a form of communication on behalf of the plant.

I pray with plants, and I will sit with plants and listen to what they tell me in the form of mental images (that little voice in your head). There have been quite a few plants that I've never met before so I'll sit with them for a while and get images and little snippets about them. The crazy part is, I've had a lot of that verified through experience. I've talked to a lot of other knowledge holders. I love it when Western science sort of catches up to Indigenous knowledge."

Kelly: "I love hearing about that intersection because so many people are skeptical one way or the other. To see some of these things coming together, it gives me hope!"

Linda Black Elk: "Yes, it's true! I worked with a chemist at North Dakota State University. He was such a wonderful man named Gary Stoltzenberg. I told him the Lakota story about when you go out to pick sand cherries, you're supposed to approach them from downwind. Otherwise, they'll smell you coming, and they'll turn bitter. Through research, he found that sand cherries have these stomata that will in fact take in animal pheromones. If they sense those pheromones, they sense a predator and start producing bitter alkaloids. It can happen very quickly, so you *are* supposed to approach them from downwind."

Meg: **"That's amazing! Are there a lot of common natural remedies that could help everyday ailments that we're just not aware of because we've lost that knowledge?"**

Linda Black Elk: "Absolutely! It's one of my favorite talks that I give: 'A Weed by Any Other Name.' These things that we call weeds, I'm so fascinated by that term. What does that imply? It's such a value-based term that really means nothing. I'm a part of a group of plant scientists on Facebook and we identify plants for people. Probably one of the most common questions we get is, 'Is this a plant or a weed?' You just can't answer that question five hundred times a day, so we usually ignore it. Sometimes those plants that people consider to be weeds in this sort of Westernized culture that we live in in the United States are some of the most amazing medicines out there. I am just flabbergasted at how the dandelion has been vilified. It is food; it is medicine. Every part of a dandelion is edible. We make cookies and bread and things out of the flowers. We make tea out of the flowers. It's delicious! We use the stems in all kinds of soups and stews and even as a spaghetti substitute. We use the leaves raw or cooked. Dandelion coffee is one of my favorite things."

Meg: "I need to try that!"

Linda Black Elk: "Dandelion in numerous studies has been used to fight cancer. It's heavy in antioxidants and I remember talking to this one gentleman. He's a physician working on a study out of the University of Windsor. They were finding that powdered dandelion root was shrinking tumors and when it wasn't shrinking tumors it was actually helping people through chemo. This is something that we spray, that we kill in our lawns, that we just consider to be something we don't want around . . . but it's this fantastic medicine! It stabilizes blood sugar for diabetics. There's the misconception, I've talked to a lot of plant scientists about this as well, that dandelions are not native. They've become so vilified as a weed that they've actually gotten this reputation that they somehow came over from Europe and that there weren't any of them before. That's totally untrue. There were always native species of dandelions here (in the United States), and I do believe that European varieties were probably brought over that have now, of course, crossed with a lot of ours. I always try to lift the dandelion up because I feel like it's been so maligned.

Another plant that I think of is stinging nettle. They are called stinging nettles for a reason but if you get your body used to them, they are amazing anti-inflammatories. We were taught that if you harvest nettles bare-handed without gloves, you'll never get arthritis in your hands. When I was growing up, I did sports, and I would always have these sore knees so I would put on a pair of shorts and run through the nettle patch so that it would sting my knees. I would be pain-free, sometimes for months, after just one treatment. I've had that happen over and over. One of my students had horrible back pain and after she took my ethnobotany class she said she did one nettle treatment on her back and had her first good night's sleep in four years. I have so many stories like that about this plant that a lot of people are afraid of, or they just ignore it. They cut it down or they compost it, but it's fantastic."

"Ancient Egyptians used stinging nettle to treat arthritis and lower back pain, while Roman troops rubbed it on themselves to help stay warm."[6]

Kelly: **"Have there been clinical trials with some of these natural remedies?"**

Linda Black Elk: "Not enough, I suppose, would be my simple answer to that question. The simple hawthorn berry is being used for heart health and even being used to heal the heart muscle post-surgery. There are now numerous clinical trials that have been done on that. There are even some cardiologists who recommend hawthorn but, because of that, you have capsules being sold in Walmart. How much hawthorn is really in those capsules? It's not that much. Someone is making bank on those and the Indigenous community that the knowledge came from has been erased. So much erasure happens in those cases and so I understand people in communities who are protective of that knowledge."

Meg: **"Is there one piece of knowledge, or a call to action, you want people to know?"**

Linda Black Elk: "In my opinion, plant medicine and traditional plant foods is the way we change the world for the better. A lot of my students say 'I used to drive my car down the road and just look out and to me it was all grass. But after you find out how diverse it is, how much food is out there, how much medicine is out there, that a plant can be used for rope, and a plant can be used to make a bow drill, you start walking on the land differently. You start being much more aware of your impact on it.' Not only does plant knowledge change the way we view the land and change the way we walk on it, but it also makes us eat differently and it makes us think."

Thank you to Linda Black Elk for this enlightening interview! We will certainly start viewing our surroundings with fresh eyes.

In 1997, the village of North Tarrytown, New York, where the story took place, officially changed its name to Sleepy Hollow. Fittingly, the Horseman is its high school's official mascot!

CHAPTER THREE
SEASON OF THE WITCH

If you mention director George Romero in a conversation, horror fans will immediately think of zombies. He basically invented the modern zombie with *Night of the Living Dead* (1968) and its subsequent sequels. A lesser-known film of his, *Season of the Witch* (1970), follows the story of Joan (Jan White) as she discovers the power of becoming a witch. Romero is said to have been inspired by the feminist movement at the time and by reading about witchcraft for another project he was working on.[1] Writer Derek Warren notes, "*Season of the Witch* is not a film for casual moviegoers. It is not even a film for all horror fanatics. It's one of Romero's most difficult films to classify, as it walks the line between drama and thriller without fully committing to any one genre. However, it is a very good movie that has finally received the release that it deserves."[2]

The film begins with sequences featuring Joan's bizarre dreams. They clearly show that she is feeling repressed and ignored in her marriage. How do our dreams help us process things in our day-to-day lives? Centuries ago, people believed that dreams were messages from beyond or divine intervention. Currently, many people use dream interpretation to understand their subconscious thoughts. What are the meanings of some of the most common dreams? If you dream that your teeth are falling out, it may mean you feel powerless in some part of your life. Dreaming that you're in an out-of-control car could mean that you feel the same about another waking situation. Having a dream that

Lucid dreaming, the ability to control your dreams, may offer therapeutic benefits by allowing individuals to rehearse actions or overcome threats during sleep.

you're falling is interpreted by some as a sign that you're hanging on too tightly to something while flying is a way for your sleeping brain to celebrate being freed from a situation.[3] In one dream, Joan sees her own reflection in a mirror, and she is gray haired and wrinkled. Even her friend Shirley (Anne Muffly) brings up her age later and her resentment about it. Why do some people have a fear of getting older? A 2018 study by the Royal Society for Public Health found that millennials viewed aging very negatively and equated it with a decline in life. The study suggested ways to address societal ageism including combining young people and old people in the same facilities, promoting age diversity in the workplace, looking at representation of older people in the media, and to end using the term "anti-aging" in skin care products.[4] When this worry about getting older starts to affect your well-being, it becomes a diagnosable condition called gerascophobia. Sufferers have frequent thoughts and anxiety about aging, it can trigger panic attacks, shortness of breath, sweating, and shaking.[5] This condition can be treated with cognitive behavioral therapy, which can help retrain the brain how to think about aging.

Joan suffers from empty nest syndrome in the film. While not a clinical condition, it is a feeling of grief and loneliness that parents may feel when their children leave home for the first time.[6]

In the movie, Gregg (Raymond Laine) suggests that people can talk themselves into being sick or getting a heart attack. Is this possible? And how does excessive worry affect our bodies? The power of belief plays a vital role in our overall well-being. Scientifically, the biochemistry of our bodies stems from our awareness. There are numerous examples in medicine of people's bodies reacting to a belief like "psychosocial dwarfism, wherein children who feel and *believe* that they are unloved, translate the perceived lack of love into depleted levels of growth hormone, in contrast to the strongly held view that growth hormone is released according to a preprogrammed schedule coded into the individual's genes!"[7]

This is directly related to the following scene when Gregg pretends to give Shirley marijuana. She starts acting high even though it's just a cigarette. What is the science behind the placebo effect? Again, our beliefs can actually control our bodies. It is mediated by the release of neurotransmitters, which impacts certain areas of the brain, and mirrors the action of pharmaceuticals or other substances on human physiology. When Shirley believed she smoked a joint, her body reacted accordingly. In 1998, a study on sixteen antidepressants found that 25 percent of the effectiveness of the drugs was due to the specific action of the drug, 25 percent was due to spontaneous remission, and 50 percent was due to the placebo effect.[8] Does this mean we should stop taking our antidepressants? No, but it proves that the mind and the power of belief are truly impressive.

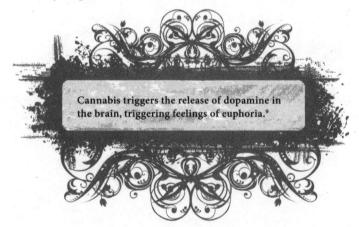

Cannabis triggers the release of dopamine in the brain, triggering feelings of euphoria.[9]

Joan accuses Gregg of being an egomaniac. What is this condition? It's a preoccupation with one's self, which may manifest in delusions of greatness. Egomania isn't considered a personality disorder but it *is* considered abnormal. Egomaniacs tend to be aggressive, vengeful, and will diminish others while lifting themselves up. They tend to lack empathy and have a sense of entitlement when it comes to their worldview. What makes people egomaniacs? The theories are split. Some experts believe it comes from a lack of love as a child, so they create their own worth by becoming self-absorbed. Others believe it stems from being overly spoiled. Regardless of how it begins, egomania can be treated through therapy if the individual is open to it.

Nikki (Joedda McClain) runs away after her mother hears her having sex in the next room. What are the statistics about teen runaways? According to the National Runaway Safeline, between 1.6–2.8 million youth run away each year in the United States.[10] Nearly half report that they ran away due to problems with their parents and may also be suffering from other issues including depression, substance abuse, or oppositional defiant disorder. Symptoms of this disorder include losing their temper often, being easily annoyed, resentful, and angry, arguing with authority figures often, and defying rules.[11] Surprisingly, 99 percent of teen runaways decide to return home, just like Nikki in *Season of the Witch*.

Joan and Gregg end up sleeping together and Gregg calls Joan "Mrs. Robinson" in reference to *The Graduate* (1967). How common are relationships between older women with younger men? The feminist movement of the 1960s is said to be a contributing factor in older women dating younger men. In fact, according to a recent AARP survey of 3,500 older singles, 34 percent of women in the forty to sixty-nine age group date younger men. And 14 percent of women ages fifty to fifty-nine say they prefer dating men in their forties or younger.[12] Sex drive may also contribute to this trend, as women reach their peak when older and men hit theirs when younger.

Joan eventually ends up accidentally shooting her husband after he unexpectedly comes home early from a business trip. Did becoming a witch help Joan accomplish everything she wanted? Perhaps. She got rid of her controlling husband, she felt powerful and beautiful, and seemed to have gained confidence. Was it witchcraft or the belief in herself? The viewer is left to decide. Critic Ruairi McCann explains:

> And though witchcraft may provide for Joan a way out, it also might ferment in her a new prison. Or an old one. And anxiety

heightened and alive in Romero's filmmaking. Which on this occasion can be rough, but what makes it fitting is that it treats witchcraft as an underground activity. An improvisation against the unconsciously monolithic social structures that Romero will subvert time and time again.[13]

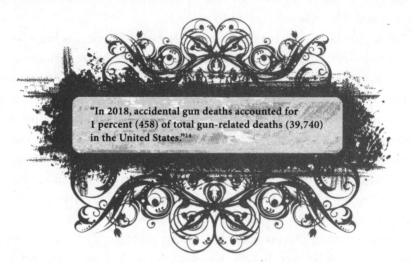

"In 2018, accidental gun deaths accounted for 1 percent (458) of total gun-related deaths (39,740) in the United States."[14]

SECTION TWO
WITCH AS OTHER

CHAPTER FOUR
WE HAVE ALWAYS LIVED IN THE CASTLE

Several years ago, I (Meg) came across a book of my father's. It was a recent edition of a few Shirley Jackson novels and short stories in one thick paperback. I had read *The Haunting of Hill House* (1959) and "The Lottery" (1948) prior to finding this new treasure, and because of my desire to expand my female horror writer's horizons, I randomly chose to read *We Have Always Lived in the Castle* (1962) next. It had an intriguing title, and an even more seductive first paragraph:

> My name is Mary Katherine Blackwood. I am eighteen years old, and I live with my sister Constance. I have often thought that with any luck at all I could have been born a werewolf, because the two middle fingers on both my hands are the same length, but I have had to be content with what I had. I dislike washing myself, and dogs, and noise. I like my sister Constance, and Richard Plantagenet, and *Amanita phalloides*, the death-cup mushroom. Everyone else in my family is dead.[1]

How could I not keep reading? Swiftly, *We Have Always Lived in the Castle* became one of my favorite novels, as Jackson's prose, tension, and atmosphere are like no other. The story of two sisters, Merricat and Constance, is a hypnotic tale about the aftermath of a family tragedy (a group poisoning that may or may not have been caused by Merricat herself). One cannot read the novel, or even watch the subsequent 2018 film, without focusing on the witch imagery and suggestion. Unlike many of the films and novels discussed in this book, *We Have Always Lived in the Castle* doesn't exhibit the typical witch tropes of broomsticks and cauldrons. It is a reflection on womanhood, and the freeing notion of

25

being singular. In both the book and film, Merricat and Constance live in isolation because of the judgments of their neighbors in a small town. This is a theme often explored in Jackson's work, as she had felt a similar ostracism when she moved to an insular community in Vermont. What is striking about this rejection is that it parallels one of Jackson's favorite topics, the Salem Witch Trials. In fact, she held such an

In the 2020 biopic, Shirley Jackson (played by Elisabeth Moss) is shown using tarot cards, a known interest. She favored the Tarot of Marseilles card set.[2]

interest in the shocking history of the trials that she wrote a nonfiction account, *The Witchcraft of Salem Village* (1956), for middle-grade readers that goes into the history of the area and paints a picture of small-town Puritan life. She wrote, "life in Salem Village was not easy at the best of times. Gaiety and merrymaking were regarded as irreligious, and the people of the village were somber and severe. Their lives were spent in hard work and religious observance."[3] This description harkens to those neighbors who have rejected the presence of Merricat and Constance,

Sixty-year-old Bridget Bishop was the first accused witch to be executed in Salem. Her most harrowing crime was believed to be the murder of her first two husbands through the means of witchcraft, of which she denied.[4]

as well as the townspeople in Jackson's iconic short story "The Lottery." The Puritans in Salem Village would surely not have appreciated women like Jackson, or her character Merricat, as they thrive on independence.

The Salem Witch Trials is a blight on our American heritage, a violent time in which women were cast out of their communities (if they were lucky) or executed by hanging. Those who were tried but not found guilty lived with a pall over their lives, as they were still regarded as others. Before the events in Salem, women accused of witchcraft had been executed on American soil, including Alice "Alse" Young in 1647. She was not formally exonerated from the crime of witchcraft until 2017!

The famous, systemic murder of innocents began in Salem in 1692 when a group of teen girls accused people in their village, mostly women, of witchcraft. Gender and race were at play, as 78 percent of those executed were women; Tituba, a house servant from the West Indies, was at the center of the accusations. Things soon got out of hand as villagers began accusing each other based on the smallest of "clues" until the jail was teeming with supposed witches. Cotton Mather, a well-known and respected minister, wrote several recommendations to the men of justice in Salem Village, including a hope that they could further their ability to suss out witches: "We cannot but, with all thankfulness, acknowledge the success which the merciful God has given unto the sedulous and assiduous endeavors of our honorable rulers, to detect the abominable witchcrafts which have been committed in the country, humbly praying, that the discovery of those mysterious and mischievous wickednesses may be perfected."[5] In the same letter, Mather made sure to include his belief that witches must be executed for their crimes against God.

Evidence in the trials often relied on witness testimony, as well as several "tests" that were considered foolproof in discovering if the accused was indeed marked by the devil. This included a search for a mole or mark on the body that was insensitive to touch. If one wasn't found, the accused was often pricked or scratched by tools until a numb place was discovered.

A bizarre form of counter-magic, the witch cake was a supernatural dessert used to identify suspected evildoers. In cases of mysterious illness or possession, witch-hunters would take a sample of the

After testifying against others, Tituba escaped execution and disappeared from Massachusetts. Her likeness has appeared in theater, novels, and TV, including *American Horror Story: Coven* (2013–2014), in which Queenie (Gabourey Sidibe) is said to be Tituba's ancestor.[6]

victim's urine, mix it with rye-meal and ashes, and bake it into a cake. This stomach-turning concoction was then fed to a dog—the "familiars," or animal helpers, of witches—in the hope that the beast would fall under its spell and reveal the name of the guilty sorcerer. During the hysteria that preceded the Salem Witch Trials, the slave Tituba famously helped prepare a witch cake to identify the person responsible for bewitching young Betty Parris and others. The brew failed to work, and Tituba's supposed knowledge of spells and folk remedies was later used as evidence against her when she was accused of being a witch.[7]

A witch cake almost sounds like arsenic in a sugar bowl! There are many links between the Salem Witch Trials and *We Have Always Lived in the Castle*. Even the literal tearing apart of the sisters' home at the end of the novel evokes the accused witches' homesteads being searched for proof of their malevolent intentions.

In 2018, fifty-three years after Jackson's death, *We Have Always Lived in the Castle* became a film. Starring Taissa Farmiga as Merricat, it follows the events of the book closely, including witch imagery. There is Jonas,

Merricat's black cat, a witch's familiar if ever there was one. She is also drawn to nature, as many varieties of witches seem to be, including in Wiccan culture. Merricat has a deep understanding of herbs and plants, so deep that she placed arsenic in a sugar bowl to kill her family. Her motive is never fully explained.

Poet, novelist, and essayist Stephanie Wytovich spoke to us about her witch lit class, as well as her own favorite witch representations in media.

Meg: **"What was your first exposure to witchcraft?"**
Stephanie Wytovich: "I grew up a very strict Catholic, but I've always loved horror and the supernatural, and like most young girls, my fascination with witchcraft started when *The Craft* came out in 1996. I mean, power, friendship, glamour, the coolest punk rock/goth clothes around? I was totally in! After watching it, I actually remember going on our family computer and praying the dial-up worked fast (yes, I'm old), just so I could look up some spells without my parents knowing what I was doing. I was always too afraid to actually try anything then, but I used to keep the spells hidden in one of my drawers because just knowing they were close was enough at that point."
Kelly: "We remember those old days of the Internet . . . "
Meg: "Would've been nice to have a spell to speed it up!"

Kelly: **"You teach a witch lit class. Could you tell us some of your favorite works to teach and discuss?"**
Stephanie Wytovich: "Sure! Some of what I've taught in the past have been: *The Crucible* (1953) by Arthur Miller, *Carrie* (1974) by Stephen King, *I, Tituba: Black Witch of Salem* (1986) by Maryse Conde, *We Have Always Lived in the Castle* (1962) by Shirley Jackson, *Chilling Adventures of Sabrina* (2018–2020) by Roberto Aguirre-Sacasa, *White Oleander* (1999) by Janet Fitch, *Circe* (2018) by Madeline Miller, and *The Ocean at the End of the Lane* (2013) by Neil Gaiman. I also teach a lot of short fiction and poetry from writers like Sylvia Plath, Anne Sexton, M. Rickert, Leonora Carrington, Carmen Maria Machado, and now Mariana Enriquez, too. When it comes to nonfiction though, the book I like to assign students to read (or encourage them to check out)

is *Waking the Witch: Reflections on Women, Magic, and Power* (2019) by Pam Grossman."

Meg: "Now I have a lot of books to add to my 'To Read' list!"

Kelly: **"In your class you teach witch archetypes like 'the hunted' and 'the fabled.' How do these help us understand the witch as she is portrayed in literature?"**

Stephanie Wytovich: "I think it's important to show different facets of the witch, so I strive to portray her from different angles throughout literature, film, and pop culture. For instance, the Owen sisters in *Practical Magic* (1995) have a different vibe to them compared to say Jezebel and her family in Eden Royce's *Root Magic* (2021), or Maleficent in various Sleeping Beauty stories/films. I think when studying the witch, we need to see her and how she's shapeshifted and been treated throughout history in order to get a full picture of the character and not a caricature based on stereotypes. Sometimes she's hunted like in Margaret Atwood's poem 'Half-Hanged Mary' (1995) and other times she's fabled, more eccentric, and dreamlike such as in Leonora Carrington's 'The Debutante' (2014). Other times she's enraged, and sometimes she's isolated, protected by the walls (literal or metaphorical) that she's put up. Furthermore, each archetype has tropes that go along with it, so when we study these portrayals, we talk about reader expectations, sure, but we also talk about how to revitalize and breathe new life into those scenarios, too, so we're not just retelling the same story about the witch, but instead giving time and energy to everything the witch can be and is."

Meg: **"I'm a Shirley Jackson fanatic. Is there power in the more subtle portrayals of witches in her work?"**

Stephanie Wytovich: "Oh, Shirley Jackson is one of my favorite literary witches, and she also created two of my favorite fictional witches: the Blackwood sisters, Constance and Merricat. I think there is so much potential and power in the more subtle portrayals of witches whether we're talking about Eleanor's abilities in *The Haunting of Hill House* (1959) or the folk imagery surrounding Jackson's infamous 1948 short story 'The Lottery.' I say this because being a witch isn't always cauldrons and crystals and spells and black cats—even though, sure, it can be those things (and Merricat might argue that is what being a witch is for her). But sometimes it's also about being independent, making your own choices, baking pies, and tending to your garden. Other times it's embracing your sexuality, helping out a friend, or not being afraid to scream and fight for what you know is right. Witches can be feral or proper, live in an old Victorian house, or hang out in a hut on chicken legs. One of the things I love most about the witch is that sometimes you don't know when you're in the presence of one. That cloak of invisibility, that ability to hide in plain sight, that's one of a witch's most powerful tools."

Kelly: **"I love that being a witch has so many unique paths! There is now a feminist reclaiming of the word 'witch.' How has this evolved over the centuries? And in recent decades?"**

Stephanie Wytovich: "The word 'witch' has certainly gone through a feminist reclaiming, as have words like 'crone' or 'hag.' I think over the past few years—politically speaking—we've seen such an attack on women's rights in addition to threats against movements surrounding Black Lives Matter and LGBTQ+ rights that the witch, in all of her rage and protectiveness, started to show her face again because she knows what it's like to be hunted, to be oppressed, to have to fight for the survival of herself and her sisters, her brothers, and her friends. She knows what it's like to be targeted, to feel unsafe, to be tortured and cast out, isolated from her people. When history repeats itself, as it always unfortunately seems to do, I think people—myself included—are taking solace in her name form, and energy in order to feel fueled, safe, and ready to fight for their rights and the rights of those around them. I think it's also important to note here that while I've been saying 'she' throughout this interview, that's more of a pronoun placeholder historically speaking; gender is no concern to the witch and anyone—man, woman, nonbinary alike—can wear the name and identity with pride. Something else to consider is that back in 1968, the W.I.T.C.H. movement was actually formed. Their name stood for Women's International Terrorist Conspiracy from Hell and they fought for women's rights, black rights, supported the anti-war movement, etc. Their members all showed up in head-to-toe black with large pointy hats and protested in the streets, using their voices to promote equality and peace. Fast forward to 2016 plus, and you have women out in the streets dressed as Handmaids, attending the women's march holding signs that say 'we are the granddaughters of the witches you couldn't burn,' or at home throwing hexes or practicing rituals to protect themselves and others against those in power. The years and faces might change, but the fight is far from over and the witch knows that. Now, thankfully, we're all starting to learn that, too, and teaming up with her has unleashed something powerful in the air, something that I hope will bring around the next era of change, one filled with equal rights, protection, and acceptance rather than actual witch hunts. With that said, wielding that name or identifier is a spell in and of itself, and a powerful one at that."

Meg: **"As a novelist and poet, how has your knowledge and research into witchcraft enhanced your writing?"**
Stephanie Wytovich: "Studying witchcraft alongside the archetype of the witch has been great for my creative process. While I've learned rituals that I sometimes perform to help conquer imposter syndrome or maybe give me a nudge in my brainstorming or drafting process, I've also learned more about history, myth, folklore, and monsters than ever before. As I often write feminist-slanted horror and dark fantasy, my knowledge of cycles, symbols, herbalism, astrology, etc. has helped to make my worlds and characters more realistic, unique, empowered . . . and dangerous (in the best of ways). I feel indebted to the witch now more so than ever as I feel her presence all throughout my work these days."

Kelly: **"What movie and/or book gets witchcraft right?"**
Stephanie Wytovich: "I love the 2017 film *Pyewacket* directed by Adam MacDonald. I actually recently watched this again after listening to a *Faculty of Horror* podcast episode on it. To me, this film touches on a lot of different things that are talked about in the witchcraft community, some of which include grief, protection, and intention. The story follows a young girl, Leah (Nicole Muñoz), who recently lost her father and is having a hard time existing in that grief alongside her mother, who is also processing the loss. Leah gets more and more interested in the occult, as it acts as a kind of catharsis and outlet for her pain, providing her with comfort, empowerment, and protection. But as the strain on her and her

mother's relationship deepens, Leah's emotions, rather than her logic, take control of the situation. Intention is such a big part of witchcraft, and unfortunately, Leah lets her rage sit in the front seat while she performs a dark ritual in the woods. Some might say that her spell worked exactly as she intended it to, and whether or not that's true, viewers learn that one, you need to know what you're doing when you walk into the craft and two, your intention, good or bad, is going to feed your spell, so you need to be 100 percent sure, without question, that you're asking for and doing what you want. This movie has some truly terrifying and chilling scenes in it, and I love that they used the name Pyewacket as the title because historically speaking, that was one of the familiars of a supposed witch that was accused by the Witchfinder General, Matthew Hopkins, during his reign in England in the late seventeenth century. Plus, it's also the cat's name in the fantastically witchy movie *Bell, Book and Candle* (1958)."

After our talk with Stephanie Wytovich, we both have a list of books and movies we are curious to check out. It was an honor speaking to someone so well-versed in witch literature, film, and history! We believe Shirley Jackson would've appreciated being included as an author in Stephanie's Witch Literature class, as she boasted of being a practicing amateur witch.

Before the witch trials, Cotton Mather wrote *Remarkable Providences* (1684), outlining his research into a thirteen-year-old girl he believed was possessed by the devil. In later centuries, doctors read his account and suggested he had described the symptoms of "clinical hysteria."[8]

CHAPTER FIVE
ʙLACKꓽSUNDAY

In the opening scene of the Italian gothic horror film *Black Sunday* (1960), there is a familiar story, both historical and cinematic. It is of Asa Vajda (Barbara Steele) tied to a stake; her fate determined by a row of men in hooded robes. It is the seventeenth century in Moldova and, like her many witch counterparts all over the world, Vajda has been sentenced to a painful demise. This fiction, unraveling before the viewer in black and white, is frightening and, for its time, rather gory.

Vajda is no repentant, sniveling witch. She is a vision of rage and confidence. As she and her lover Javutich (Arturo Dominici) are moments from death, she is quick to curse those who have wronged her, promising that she will be back to exact her revenge on her brother's descendants, as he was the one to rat her out.

Instead of being consumed by flames on the pyre, Vajda and Javutich are killed with what can only be described as a crude torture device. A mask with spikes is placed over their faces and then pounded with a mallet until they die. It is most similar to an iron maiden, a human-shaped contraption filled with sharp spikes that impale those unlucky to be inside. In the modern age we are used to imagery of this casket of spikes when watching films set in the Middle Ages. In Roger Corman's 1961 version of the Edgar Allan Poe story *The Pit and the Pendulum*, Elizabeth (also played by *Black Sunday*'s Barbara Steele) pretends to die

The classic design of the iron maiden was perfected in Nuremberg, Germany. It consists of a feminine, seven-foot-tall form with a symbol of the Virgin Mary affixed to the top.[1]

in an Iron Maiden to drive her brother (Vincent Price) insane. Ichabod Crane (Johnny Depp) in *Sleepy Hollow* (1999) is also plagued with visions of his mother's (Lisa Marie's) death within this horrific device. Vajda and Javutich are fortunate to endure just the mask in the film, as the real iron maiden would prolong death:

> Once inside, the doors were shut on the victim and the spikes would pierce several organs of the body. The spikes were supposedly short and positioned so that the victim wouldn't die quickly. This meant that they would only result in relatively small wounds and the victim would bleed to death over the course of several hours. To add to the abject horror of it all, two spikes were positioned specifically to penetrate the eyes.[2]

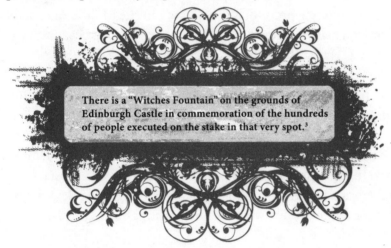

There is a "Witches Fountain" on the grounds of Edinburgh Castle in commemoration of the hundreds of people executed on the stake in that very spot.[3]

In the late 1700s, German philosopher Johann Philipp Siebenkees wrote the first historical description of the torture device in regard to a 1515 death of a coin forger. This led to a macabre interest, causing many to be manufactured for museums in Europe and the US. These replications of the supposed real thing came with vague, long-ago stories of people dying within their sharpened walls.

Before Siebenkees, there were other mentions of similar torture devices, like by Greek historian Polybius who lived in 100 BC. Journalist Stephanie Pappas outlined his tale in her article for *Live Science*:

> Polybius claimed that the Spartan tyrant Nabis constructed a mechanical likeness of his wife Apega. When a citizen refused to

pay his taxes, Nabis would have the faux wife wheeled out. "When the man offered her his hand, he made the woman rise from her chair and taking her in his arms drew her gradually to his bosom,' Polybius wrote. 'Both her arms and hands as well as her breasts were covered with iron nails so that when Nabis rested his hands on her back and then by means of certain springs drew his victim towards her, he made the man thus embraced say anything and everything. Indeed, by this means he killed a considerable number of those who denied him money."[4]

Both Polybius and Johann Philipp Siebenkees penned intriguing tales, yet there is no concrete proof that those living in the Middle Ages actually used the iron maiden to torture and kill. Medieval expert Peter Konieczny further asserts that these claims are flights of fancy:

It is likely that Siebenkees just invented this story, but by the early nineteenth century the iron maiden was being displayed in Nuremberg and other places. One of them was even exhibited at the World's Fair in Chicago in 1893, which furthered its reputation. Even though the iron maiden of Nuremberg was deemed a fake, it still has a reputation of being a real medieval torture device, one that some books claim was used as far back as the twelfth century."[5]

Don't let the mystery surrounding the iron maiden give you the warm fuzzies! Just because that particular casket of thick nails may be more fiction than fact, there was plenty of torture doled out to those accused of witchcraft. In Scotland, an estimated fifteen hundred people, mostly women, were executed for "sorcery" over the span of the fifteenth through the eighteenth centuries. Most were burned and strangled, but not before withstanding painful indignities. Like Asa Vajda in *Black Sunday*, some of those accused in Scotland were forced to wear a mask. Known as a Scot's bridle or branks, these muzzled masks were outfitted with a padlock and pieces of iron that held their tongue flat, rendering them unable to speak.

Scotland holds a long and complicated history with witchcraft, as evidenced in the example of Scottish "witch" Helen Duncan. Born at the end of the nineteenth century, Duncan lived in Callender and was known as a medium who would conduct séances in order to convene with the dead. Her point of fame was that she was adept at producing and swallowing ectoplasm, a supernatural slime that would aid in her communication. "In 1931, however, the London Spiritualist Alliance revealed it to be no more than an amalgam of cheesecloth, paper, egg white—and toilet roll."[6]

Ectoplasm is most often synonymous with *Ghostbusters* (1984) as the substance sucked up in their proton packs. Slimer was absolutely dripping with it! In reality, there is a scientific definition for ectoplasm: the outer layer of cytoplasm of an amoeba.

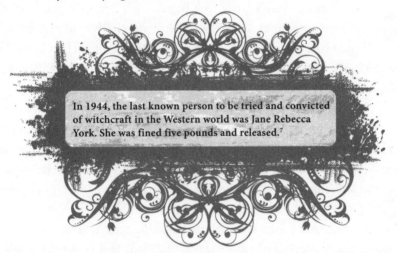

In 1944, the last known person to be tried and convicted of witchcraft in the Western world was Jane Rebecca York. She was fined five pounds and released.[7]

The Witchcraft Act of 1735 strove to eradicate the belief in witchcraft once and for all among educated people, the judiciary, and the Anglican Church. Its passage meant that it was no longer possible to be prosecuted as a witch in an English or Scottish court. It was, however, possible to be prosecuted for pretending to "exercise or use any kind of witchcraft, sorcery, enchantment, or conjuration, or undertake to tell fortunes." Supposed contact with spirits fell into this category. Some two hundred years later, when this act was finally repealed, it was pronounced to have been "a most enlightened measure, well in advance of public opinion" by the then home secretary, James Ed.[8]

Because the Witchcraft Act was still relevant during Helen Duncan's lifetime, she was apprehended in 1933, as well as in 1944, when officials busted through her home during a séance. This time she was not only jailed but also put on trial for offenses under the Witchcraft Act. The UK newspapers had a heyday with Duncan's weeklong trial, often sensationalizing her depiction as accused witch through cartoon depictions of Duncan as a stereotypical witch on a broomstick. She was sentenced to nine months at the Holloway Prison in London, and despite this setback, was known to continue her popular séances up until her death in 1956.

Italian filmmaker Mario Bava, often mentioned in the same breath as *Suspiria*'s (1977) Dario Argento (see Chapter 15 on page 115), was known for his low budget, visually stunning horror films. *Black Sunday*, Bava's first movie as director, solidified his talents and gave both him and ingenue Barbara Steele their first brush with fame. Steele, a British actress, went on to star in a number of Italian, British, and American horror films. Most recently the eighty-three-year-old actress lends her voice to *Castlevania* (2017–2021). I (Meg) for one, remember her most vividly as Dr. Mengers in 1978's *Piranha*.

Inspired by the novella *Viy* (1835) by Russian author Nikolai Gogol, Mario Bava adapted Gogol's tale into the film's script. Set in Ukraine, *Viy* tells the story of three students who come across an evil witch in the gloom of the woods. She has the ability to make the men see hallucinations, and she, herself, can shapeshift. Because of many script changes, Bava's adaptation became muddied, and its resemblance to *Viy* is passing at best. But the vital piece, the evil, unrelenting witch, remains.

One unique aspect of *Black Sunday*'s Asa Vajda is that she is vampiric. This combination of both witch and vampire makes Asa Vajda a formidable monster. Just like the film, true fear of vampires was as palpable to the people of Europe as the distrust of witches. This is evidenced in a 2014 archeological finding in Poland: a skeleton with a brick in its mouth and stakes driven through the legs. Known as an "anti-vampire" burial, these tortures were done to ensure that the dead would not rise like Asa Vajda and her lover, hungry for blood and revenge. While this sounds like a medieval frame of mind, scientists at Poland's Pomeranian University verified that the woman had been killed in the eighteenth century! Through DNA sequencing, they were also able to discover that she had blonde hair, blue eyes, and was at least sixty-five years old at the time of her death. Most disturbingly, the scientists said "the injuries to the body were sustained while she was still alive and that she likely died from them. They believe she was accused of being a witch, then tortured—with the leg piercing being part of this. Her accusers are believed to have placed a brick in her mouth to ensure 'protection' from her black magic—even after her death."[9] (This echoes the iron maiden-esque mask on Asa Vajda, as it was secured to her face after death to keep her evil from rising. Too bad for the Ukrainian villagers in *Black Sunday* that curious Dr. Choma Kruvajan (Andrea Checchi) not only removes her mask, but revives her!)

The body found in Poland is one of many "anti-vampire" burials. Anthropologist Matteo Borrini from the University of Florence explains the thinking that led to these bizarre burials: "Gravediggers reopening mass graves would sometimes come across bodies bloated by gas, with hair still growing, and blood seeping from their mouths, and believe them to be still alive. The shrouds used to cover the faces of the dead were often decayed by bacteria in the mouth, revealing the corpse's teeth, and vampires became known as 'shroud-eaters.'"[10] According to medieval medical and religious texts, the "undead" were believed to spread pestilence in order to suck the remaining life from corpses until they acquired the strength to return to the streets again. "To kill the vampire, you had to remove the shroud from its mouth, which was its food, like the milk of a child, and put something uneatable in there."[11]

One disturbing "anti-vampire" burial was that of a ten-year-old child in the fifth century, who scientists were able to determine died of malaria.[12]

It's clear from both folklore as well as science that the fear of witches was a reality across Europe. What happened in our own Salem, Massachusetts, was merely an echo of our ancestors' misunderstanding of disease. Like Asa Vajda coming back to destroy her own descendants, this cycle of torture and abuse against those who are merely different would continue if not for the advances of science that have proven there is nothing to fear.

CHAPTER SIX
THE MANOR

In *The Manor*, widow and grandmother Judith (Barbara Hershey) moves into an assisted living home after a stroke and the diagnosis of Parkinson's disease. This transition into a stately manor with gothic architecture is what horror movies are made of, as the atmosphere immediately gives both Judith and the viewer a sense of dread.

As we come to learn, residents are dying mysteriously in *The Manor*. Judith soon witnesses a terrifying creature in her bedroom, a monster that lurks in the shadows, its limbs and skin borne from a tree. This depiction of a "tree monster" is rooted (ha!) in mythology that was made popular by J. R. R Tolkien's Ents in his 1954 novel *The Lord of the Rings* and has appeared in everything from the TV series *Sleepy Hollow* (2013–2017) to the clinging branches inhabited by demons in *The Evil Dead* (1981).

In Greek mythology, dryads were nymphs that lived inside trees. There were different dryads inhabiting various species of tree. For instance, the Melai dwelled in ash trees and were believed to be formed from the spilled blood of Uranus.[1]

The tree itself, a true royal in the hierarchy of nature, naturally brings about the connotation of pagan symbolism. Being a witch has an inherent connection to nature. From hoodoo to Wicca, plants are a foundation of magic. This monster, however, seems to be preying on the weak, not developing a peaceful love of the outdoors. One story out of Africa may have planted (haha!) the notion of evil trees into cultural relevance. In 1881, Carl Liche, a German explorer, came back from Madagascar with a wild tale about a living tree that the Mkodo tribe sacrificed young women to:

From the top of the tree sprout long hairy green tendrils and a set of tentacles, constantly and vigorously in motion, with . . . a subtle, sinuous, silent throbbing against the air. [Presented a woman as offering], the slender delicate palpi, with the fury of starved serpents, quivered a moment over her head, then as if instinct with demoniac intelligence fastened upon her in sudden coils round and round her neck and arms; then while her awful screams and yet more awful laughter rose wildly to be instantly strangled down again into a gurgling moan, the tendrils one after another, like great green serpents, with brutal energy and infernal rapidity, rose, retracted themselves, and wrapped her about in fold after fold, ever tightening with cruel swiftness and savage tenacity of anacondas fastening upon their prey."[2]

The "Triple Goddess" who has roots in Paganism and is most likely inspired by Hecate of Greek mythology, is defined by her three states of being: Maiden, Mother, and Crone. These represent a woman's life stages, with Crone being the wisest.[3]

This unbelievable account was corroborated by American explorer Chase Osborn who wrote a book on the subject, *Madagascar; Land of the Man-Eating Tree* (1924), within which he added that other tribes on the island were aware of the blood-hungry tree. It wasn't until 1955 that the entirety of the story was exposed to be a hoax by science writer Willy Ley. The tribe, the tree, and even German explorer Carl Liche were all fabricated. Despite it being false, the story resonated and was perhaps the inspiration for such murderous creatures as Audrey II from *The Little Shop of Horrors* (1960).

Judith comes to learn that her fellow residents are using witchcraft to slow, or outright stop, their aging. This naturally comes with a cost: sacrificing innocents. Like the women of *Death Becomes Her* (1992), Judith is desperate to retain both her body and mind. While crones may be wise, she is not prepared for the loss of her former life.

We had the honor of speaking with the writer and director of *The Manor*, Axelle Carolyn, about her film!

Kelly: **"First, can you tell us about the witches that left an impression on you in childhood?"**
Axelle Carolyn: "I guess my earliest memory is probably Gargamel, the warlock from *The Smurfs* (1981–1989), haha! Bavmorda (Jean Marsh) from *Willow* (1988) also made a big impression. I loved scary witches as a kid. And *Bedknobs and Broomsticks* (1971)! As a teen, I'd have to mention the cooler, more relatable witches from *The Craft* and *Buffy the Vampire Slayer* (1997–2003). Although, come to think of it, in both stories they get punished for abusing their powers, which doesn't make them the tales of empowerment I wish they were!"

Meg: **"Good point! *The Manor* takes place in an assisted living home. What made you want to tell a story centering on older characters? Do you feel they are largely ignored in the film or horror genre in general?"**
Axelle Carolyn: "They certainly are. And women in particular. The way society ignores them, and the film industry by and large depicts them, could give you the feeling that women lose their worth when they're no longer of child-bearing age, which is as insulting as it is absurd. There are so many women in my life of Judith's age and they're sexy, charismatic, strong, and hilarious. There are plenty of older women in horror movies, but they're usually horrifying, hence the term hagsploitation (ugh). I wanted an aspirational lead character. It's important to have role models who make you feel okay about getting older. The movie used to be called *Crones*. I love the fact that the term 'crone' used to mean a woman of

wisdom, power, and magical abilities, and nowadays it just means old and ugly. To me, it perfectly encapsulates the shift in how Western society views the elderly."

Kelly: **"Wow! I didn't know the meaning has shifted over the years. Did you research a particular branch of witchcraft when writing this film? What influences and knowledge did you already come with, and what did you need to learn?"**

Axelle Carolyn: "I grew up in Europe, and I think druids, the green man, and Celtic culture in general seeped into my brain. I'd originally envisioned the story with moody skies and dark forests, and much more nature-centric than it ended up being, but as we were shooting the movie as part of a series, we had to shoot in Los Angeles, so the look turned out quite different. The tree played a bigger part in the original script, it was supposed to be a massive oak tree. In the end, the only interesting one we found was this incredible tree which had been struck by lightning, so its trunk was dead and hollow, yet one half kept growing and having perfectly healthy leaves. Pretty symbolic, I'd say! I love the idea of witches

There were many reputed witches in Ireland, like Alice Kyteler, who in the thirteenth century outlived her four husbands, inheriting their wealth. This caught the attention of the Bishop of Ossery, as well as gossip that Kyteler practiced satanic rituals, causing a witch hunt. Kyteler escaped to England.[4]

drawing their power from nature. Of course, what they're doing in the film is a perversion—they're cheating nature, and as one of them says, cheating nature always comes at a cost, so they end up having to sacrifice lives in the process. But using the cycle of life and nature to go against the established order of things seemed fun."

Meg: **"The creature was terrifying! Were there any creature influences? Could you tell our readers what directing those sequences was like?"**
Axelle Carolyn: "Thank you! We called him the Minion, because he does the witches' bidding. He's part of the tree they draw their power from, so he had to be made of branches and bark. The designer, Martin Astles from Illusion Industries, put together all these beautiful texture references, and I'd brought a lot of visual references, too, for inspiration. I didn't want anything too demonic looking. Instead, he's more like a skeleton with these big human teeth and branches that form ribs and a spine. The creature performer, Mark Steger, really made him unique through the way he crawled and moved. Filming those scenes was my favorite part of the shoot. The first day with him was on Halloween, and I was pretty much giddy with excitement seeing our creature come to life."
Meg: "Ah! That sounds like so much fun! Perfect Halloween!"

Kelly: **"You wrote for *Chilling Adventures of Sabrina* (2018–2020). Were there any witchy inspirations that moved you to write *The Manor*? Do you think these witches could exist in the same universe?"**
Axelle Carolyn: "They probably could coexist, although Sabrina presupposes that witches are born, and in *The Manor* everyday people are able to make the choice. But Roberto Aguirre-Sacasa, the showrunner, wanted to explore different types of witchcraft and magic, so it's a very inclusive world! I was a staff writer on the whole first season, and it was really fun to spend six months with these super talented writers, discussing witchcraft and the devil every day.

As for inspiration for the feature, there's a number of movies that played on my mind, but I guess the main one was *Rosemary's Baby* (1968), in that it doesn't focus on the supernatural as much as on the fact that its protagonist can't trust the people around her. That idea of gaslighting, of your loved ones not being able to help, of the people who are supposed to help you being the ones who turn against you—I find that much more terrifying than any creature or supernatural threat. I wanted to stay clear of any devil or satanic reference, however. These witches' magic predates Christianity. For example, the scene is pretty quick so you can't quite tell, but when they put up Christmas decorations, they're all pagan and nature-based—no baby Jesus and no Santa."

Meg: **"I was obsessed with Judith's horror-centric wardrobe and interests! Do you hope to be like her when you reach that age? (I do . . .)"**
Axelle Carolyn: "Oh, absolutely. She was 100 percent written to be the person I hope to become; fun, tongue-in-cheek, independent, potty-mouthed, strong-headed, and effortlessly cool. She doesn't take shit from anyone. There's this idea that at a certain age, women turn into Angela Lansbury—cute little grandmas who bake cookies and are always very proper—but that'll never be me. Her horror-centric tastes were also an easy way to give her something in common with her grandson; the way she dresses is kind of a more evolved version of his teen horror nerd attire."

Actor Mark Steger who brings the tree creature "Minion" alive in *The Manor* has also brought us the terrifying movements of the Demogorgon in *Stranger Things* (2016–) as well as "Scarecrow" and "Pale Lady" in *Scary Stories to Tell in the Dark* (2019).[5]

Kelly: **"Finally, tell us what you're working on! What are your upcoming projects? Any more witches in the future?"**

Axelle Carolyn: "Definitely more witches! Who knows what the future holds, and which project will see the light of day first, but I have a bunch of witch-centric ideas, so it's a pretty safe bet!"

It was so awesome being able to talk to super talented filmmaker Axelle Carolyn about witches, tree creatures, and of course . . . *The Smurfs!*

CHAPTER SEVEN
EVE'S BAYOU

From cackling hags to sirens who are purveyors of vicious magic, witches populate horror media with prolific abandon. They are often filled with hate and menace, their spindly fingers extending into all sorts of horrific misadventures. *Eve's Bayou* (1997) is *not* that kind of movie. Considered a Southern gothic drama, *Eve's Bayou* evokes different feelings for the viewer than a horror film. "Southern Gothic is a literary style that takes gothic themes and places them in a magical realist American South setting. It's a genre characterized by contrast, mixing elements of dark romance, horror, and the supernatural."[1] Flannery O'Connor, William Faulkner, and Zora Neale Hurston are known for their Southern gothic novels. My (Meg) favorite example of the genre is the haunting book by Toni Morrison, *Beloved* (1987), which was turned into a film starring Oprah Winfrey in 1998. This subgenre of the gothic literary movement naturally transitioned to film. Some swayed more toward the dramatic side like the Tennessee Williams's play adaptation of *Cat on a Hot Tin Roof*

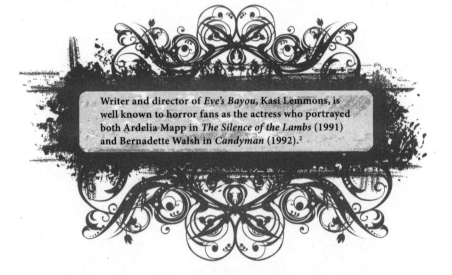

Writer and director of *Eve's Bayou*, Kasi Lemmons, is well known to horror fans as the actress who portrayed both Ardelia Mapp in *The Silence of the Lambs* (1991) and Bernadette Walsh in *Candyman* (1992).[2]

(1958) or *The Beguiled* (1971), while others leaned into the horror and supernatural like *The Skeleton Key* (2005) about a hospice aide thrown into the world of hoodoo, or *Frailty* (2001), in which a father is convinced his sons are inhabited by demons (Bill Paxton's most gut-wrenching performance!).

Eve's Bayou is an atmospheric drama that stands somewhere in the middle. The horror of this film is rooted in reality (adultery, violence) while the supernatural elements are often the only comfort and control the female characters have in the male dominated domain of 1960s rural Louisiana.

Eve Batiste (Jurnee Smollett) is a ten-year-old girl who, unlike other fledgling witches in many a film, does not question her special abilities. Like her Aunt Mozelle (Debbi Morgan), Eve is able to see the future through premonitory, or precognitive, dreams. Early in the movie, she wakes after having a nightmare about her Uncle Harry. Through black and white flashes, we watch her dream unfold; Harry dies in a car accident. As soon as she opens her eyes we hear the dreaded ring of the phone in the middle of the night, and her mother, Roz's (Lynn Whitfield's) anguished shouts. The next scene confirms that Harry did

die as Eve had witnessed in her sleep. In the media, witches are often shown to have this ability, also known as oneiromancy, as seen in *Eve's Bayou*, as well as in shows like *Witches of East End* (2013–2014). The power of peering into what has yet to come is of course demonstrated in other witch ventures like tarot card and tea leaf reading. Yet, seeing the future within a dream, a rather uncontrolled psychological force, is something many nonwitches claim to have experienced. Scientists, from psychologists to neurologists, have attempted to understand this phenomenon.

Shortly before his assassination, Abraham Lincoln told of a prophetic dream in which he looked upon his corpse in the East Room of the White House. This is exactly where his casket was placed about two weeks later.[3]

When aviator Charles Lindbergh and his wife Clara awoke to find their baby missing due to a brazen kidnapping in the spring of 1932, America, and the entire world, fixated on the tragic events that followed. The widespread interest in the Lindbergh kidnapping inspired Harvard psychology professor Henry Murray to study precognitive dreams. A few days after Charles Jr. was plucked from his crib, Murray persuaded a national newspaper to run an article asking readers to submit a letter to Murray if they had had a dream that in some way predicted the child's fate. He received about thirteen hundred letters. Though, it wasn't until after the finding of poor child's corpse, and the apprehension of his murderer, Bruno Richard Hauptmann, two years later, that Murray delved into the letters.

> Case closed; Murray set to work. He examined his collection of alleged premonitions for three important pieces of information that would have helped the police investigation enormously—the fact that the baby was dead, buried in a grave, and that the grave was near some trees. Only about 5 percent of the responses suggested that the baby was dead, and only four of the thirteen hundred responses mentioned that he was buried in a grave near some trees. In addition, none of them mentioned the ladder, extortion notes, or ransom money. Exactly as predicted by the 'dream premonitions are the work of normal, not paranormal forces' brigade, the respondents' premonitions were all over the place, with only a handful of them containing information that subsequently proved to be accurate. Murray was forced to conclude that his findings did 'not support the contention that distant events and dreams are causally related.'[4]

In a more recent study (2005) than Murray's Lindbergh experiment, researchers in the UK asked a group of 386 participants about their experience with premonitory dreams. They also inquired about the participants' belief in the paranormal to understand if there is an affirmation bias involved in one's connection from dream to future event:

> For example, imagine that a person holds a belief that left-handed people are more creative than right-handed people. Whenever this person encounters a person that is both left-handed and creative,

they place greater importance on this 'evidence' that supports what they already believe. This individual might even seek proof that further backs up this belief while discounting examples that don't support the idea."[5]

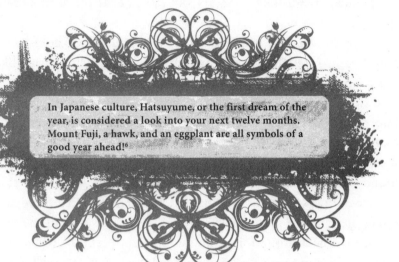

In Japanese culture, Hatsuyume, or the first dream of the year, is considered a look into your next twelve months. Mount Fuji, a hawk, and an eggplant are all symbols of a good year ahead![6]

The study found that 46.3 percent claimed to have had a precognitive dream, and that about 42 percent of those participants attributed their dream to the paranormal rather than to chance. Interestingly, they found no age or gender correlations for those who believed their dream manifested from the supernatural, which contradicts past studies in which women were said to be more prone to believing.

An affirmation bias was discovered in the 42 percent by asking three questions (example: is your back hurting right now?), of which those in this category were most likely to answer yes. This, of course, is a simplistic look at the data, as the researchers recognize that there are many factors at play:

This finding of an affirmative bias may indicate the presence of processes such as suggestibility or inclusive categorization; the latter accords with Bressan's (2002) view of paranormal belief as related to a propensity to connect events rather than to underestimate the likelihood of their occurrence. This finding may also have implications for the accuracy of self-report measures of paranormal experience and belief.[7]

There was also data to suggest that those more educated were less likely to link the paranormal to a dream, but there needed to be more research to prove causation.

Okay, so researchers aren't exactly giving us much hope to believe that dreams foretelling the future are from our witch powers, like Eve in *Eve's Bayou*. But oneiromancy has been around since, well, the Bible! God often spoke to and warned his disciples in their dreams. Dream interpretation has long been a part of literature, the first example being that of Sumerians in Mesopotamia in around 3100 BCE. They recorded their dreams and used them as divination. Even Sigmund Freud has found our dreams to hold psychological resonance.

Dreaming is just one way that witches can attain knowledge about the future (and the past and present, too!). The use of cards to read one's fortune has been around since the late sixteenth century, when cards for a parlor game in Europe were used to tell the future. It wasn't until 1791 that the first tarot card deck for the purpose of divination was created by French occultist Jean-Baptiste Alliette. Eventually, the rise of spiritualism in the Victorian era led many interested middle- and upper-class families to purchase their own tarot deck. While the first decks were inspired by Kabbalistic imagery, there are now hundreds, if not thousands, of different styles to choose from. My (Meg) deck is adorned with crows, some wearing crowns and others holding swords in their wings!

One form of fortune-telling that I'd never seen before *Eve's Bayou* is bone reading. Mozelle handles a small bag of animal bones. She spills the contents of the bag on the table and, after interpreting the placement of the bones, is able to see visions of what her clients are seeking. This practice, also known as scapulimancy, was favored in the Shang Dynasty in ancient China, where shoulder blades of cattle and turtle shells were used to seek enlightenment:

Before divination could be performed, the bones had to be prepared. First, they were stripped of flesh. Then, craftsmen drilled a series of small pits into the shells. When it was time to ask the spirits a question, they inserted hot brands into the hollows and waited for the shell to crack. The shape of the crack contained the spirits' response to each question. The questions were inscribed onto the surface of the bone. The questions covered the full range from personal to political. The diviners of the Shang court asked about the state of the harvest, the success of military campaigns, and the proper order of sacrifices, including human sacrifices.[8]

Most likely, Mozelle, a native of Louisiana, came to know this skill described as a bone basket:

> Often, bones are mixed in with other items—shells, stones, coins, feathers, etc.—and placed in a basket, bowl, or pouch. They are then shaken out onto a mat or into a delineated circle, and the images are read. This is a practice found in some American hoodoo traditions, as well as in African and Asian magical systems. Like all divination, a lot of this process is intuitive, and has to do with reading the messages from the universe or from the divine that your mind presents to you, rather than from something you've got marked down on a chart.[9]

The traditional practices of hoodoo are seen throughout *Eve's Bayou*. One example is a remedy Mozelle prescribes for bankruptcy. She instructs that a skin of chamois be filled with a piece of lodestone, John the Conqueror root, tied with the devil's shoestring, and then covered with five drops of holy oil. This sachet must be kept close to the skin.

A visit to popular online hoodoo supply shop Art of the Root led me to find many different oils that are commonly used in spells. There was no "holy oil" but there was "Holy Spirit oil," which the purchaser was suggested to drip in a glass jar with a written wish inside. They also carry John the Conqueror oil, as well as a candle with the root that purports to bring the luck of money! There is also lodestone oil, which again is used to manifest fortune. Even though Mozelle is a fictional witch, her concoction is rooted in authentic hoodoo.

Another online shop, The Witch Depot, carries the devil's shoestring, also known as ribbon grass. It's a plant that grows mainly in Texas, described by the shop as "widely believed to have the power to protect against evil, harm, or gossip. Some folks carry them in a red flannel bag for protection from crossed conditions, others carry them in a green flannel bag for money luck."[10]

Eve's Bayou is a Southern gothic film with both real and supernatural ramifications. Both worlds twist together in Eve and her family's lives, complicating their outlook on the future. The film's subtle but integral use of hoodoo solidifies it as a must-watch witch film.

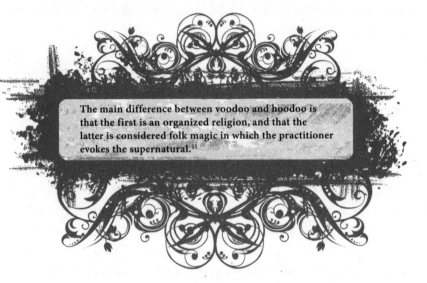

The main difference between voodoo and hoodoo is that the first is an organized religion, and that the latter is considered folk magic in which the practitioner evokes the supernatural.[11]

SECTION THREE
FAIRY TALES

CHAPTER EIGHT
GRETEL AND HANSEL

We probably all heard the fairy tale "Hansel and Gretel" (1812) by the Brothers Grimm when we were young. A brother and sister are taken in by a cannibalistic witch and are fattened up in order to become the feast. They ultimately kill the witch, escape, and live happily ever after with her riches. Folklorists have differing opinions on the lesson and meanings behind the story including oral greed, the symbolic order of a patriarchal home, and the denigration of the female characters.[1] Whatever the interpretation, the tale makes a lasting impression on all who hear it.

There have been numerous adaptations on both stage and screen of this classic children's story. A recent one, *Gretel and Hansel* (2020), offers another look into this food-filled dwelling that the witch calls home. Director Oz Perkins purposely flipped the names in the title in order to put more focus on the character of Gretel. In an interview he stated, "There was more of a feeling like Gretel having to take Hansel around everywhere she goes, and how that can impede one's own evolution, how our attachments and the things that we love can sometimes get in the way of our growth."[2]

> "Jacob Grimm's work led to a more rigorous, scientific approach in historical linguistics, which ultimately led the way to modern formal linguistics as a science."[3]

Gretel is essentially put into the role of raising her brother, Hansel, in this film. Siblings raising siblings happens every day for various reasons. Death, illness, addiction, a prison sentence, and complicated work schedules can all contribute to older siblings having to take on the duties that a parent usually would. How does this affect children? For the older sibling, they are forced to grow up immediately and may experience some culture shock. The stages of culture shock are honeymoon, crisis, adjustment, and acceptance. In the honeymoon stage, everything may seem fine and kind of fun. But at some point, it becomes too much and starts to feel overwhelming. This is the crisis stage. As the sibling becomes more used to their new duties and responsibilities, they enter the adjustment stage and work things out. Finally, when they're completely used to taking on the parenting role, they reach the acceptance stage.

The nostalgic, eighties-sounding score by Robin Coudert adds depth and atmosphere to the beautifully shot film. About his unique and haunting score Coudert noted, "the idea was to avoid the traditional musical schemes used in tales—such as the use of symphonic orchestras—and therefore find a more original and specific color. I find it essential to create melodies that we can sing or whistle as, in horror cinema, it is usually the opposite, where the music rather has a tendency toward structure and abstraction. For this project, which is a film about kids, it seemed important to have that."[4] *Gretel and Hansel* is a slow burn type of film that builds dread with its pacing.

Triangles appear throughout *Gretel and Hansel* in the set decoration, especially in reference to the witch. What significance and symbolism do various shapes hold throughout history? Pythagoras, a Greek mathematician and philosopher in the sixth century BCE, believed that geometry could offer a rational understanding of God, man, and

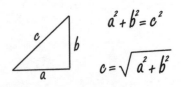

$$a^2 + b^2 = c^2$$

$$c = \sqrt{a^2 + b^2}$$

The Pythagorean theorem is used in a line in The Wizard of Oz (1939) to prove that Scarecrow has a brain.

nature. He taught that all things are numbers, including music and art, and that math was spiritual. Pentagrams are used in Wicca as a symbol of protection and warning while Pythagoras thought they represented the five elements that make up humans: fire, water, air, earth, and spirit.[5]

Pythagoras is best known for the Pythagorean Theorem which states A squared plus B squared equals C squared.

The duo, when at their hungriest, discover mushrooms in the forest and decide to eat them. Unfortunately, they cause hallucinations. How do you know if a mushroom is safe to eat? A basic tip that can be helpful is to choose mushrooms without white gills. While some mushrooms with white gills are edible, the most poisonous ones almost always have them. Look instead for mushrooms with brown or tan gills. Like the mushrooms in the film, if the cap or stem is red, you should avoid eating them. Their color is a natural warning system to stay away! You should also avoid eating mushrooms that have scales or spots on them and avoid ones that have a ring around the stem.[6] The type of mushrooms that Gretel and Hansel consumed were likely psilocybin mushrooms which contain a hallucinogenic. The effects can be felt thirty minutes to an hour after being ingested and will not only cause visual and auditory hallucinations but also emotional reactions. Err on the side of caution, friends, and don't eat wild mushrooms without proper training!

Director Oz Perkins was inspired for this film by the thought that we eat our feelings. What is the science behind this? Emotional eating is commonplace for a lot of people. You may eat to relieve stress, soothe depression, or to reward yourself. It becomes a problem if or when it is the sole coping mechanism for day-to-day emotions. People who are emotional eaters fall into a cycle of eating too much, feeling guilty about it, and then soothing their guilt with more eating. Experts recommend becoming aware of the difference between physical hunger and emotional hunger, avoiding triggers like having certain foods around, and eating mindfully to feel when you're full. Some easy habits to incorporate are to keep a food diary to track patterns of emotional eating, waiting five minutes before eating to see if the feeling will pass, and finding an alternate activity to cope with a strong emotion other than eating.[7]

The witch is a gorgon (you can read more about gorgons in our book, *The Science of Women in Horror*). Typically, we would feel safe with this older, motherly type, but she is a villain. She says she took on the form of an older woman to appear kind and weak. But truly, she is powerful. The witch is able to sense that a storm is coming. Is this a spooky power that she possesses, or can some people predict the weather? They can! Research has shown that bones or joints that are weakened by age or injury are, in fact, sensitive to changes in barometric pressure. When barometric pressure drops, it signals a change in weather and can induce physical pain in people with those weakened joints or bones. What is physically happening inside the body to trigger this response? Due to the invasiveness a possible study would need to include, only hypotheses have been made. Using anecdotal evidence, researchers believe air pressure changes affect the body's fluid levels around joints.[8] Ask for forgiveness from your grandma if she ever predicted the weather by her aching knee and you doubted her!

Hansel experiences too much of a good thing by having an abundance of food and it seems that he's finding it all a bit boring. What's the psychology behind this? Steve Glaveski said to think of all things in life like this: "At what point does the value I derive from this start to subside or hurt other aspects of my life?"[9] Too much of a good thing can definitely go overboard. For example, consuming too much caffeine can have a negative effect on your body and concentration levels even if the original intent was to give you some extra pep. Planning is another example. If you plan too much for something, you may be putting off getting started, or you may be missing out on some creative or spontaneous journeys by not being open-minded enough about what you're planning. The trick is to find some balance. Perhaps if Hansel had paced himself with the food consumption or mixed up the variety of foods he was eating, he wouldn't have gotten sick of it so quickly.

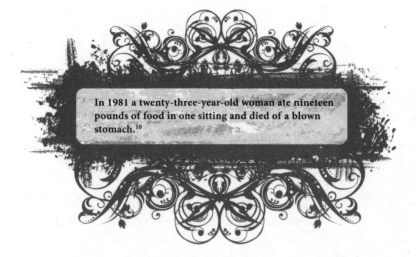

In 1981 a twenty-three-year-old woman ate nineteen pounds of food in one sitting and died of a blown stomach.[10]

One of the most impressive scenes in *Gretel and Hansel* takes place at the dinner table. The witch pulls a long braid of hair out of her mouth after eating a bite of food. It was all done in one shot, in one take, using practical effects. Before this, the witch tells Gretel that she must become immune to poison by ingesting it. Could it be possible to build up an immunity to something dangerous, akin to the famous poisoned drink scene in *A Princess Bride* (1987)? There is a practice named mithridatism that involves slowly giving oneself small amounts of poison so as to develop an immunity to it. It is named after Mithridates VI, the King of Pontus, who did this very thing because he was so paranoid of being poisoned by his mother. He had reason to be suspicious, though. Apparently, she killed his father by poisoning him and ordered for her son to be killed the same way! Mithridates fled after losing a battle and, ironically, tried to die by suicide by ingesting poison but his body was immune to it. He died instead by sword at his own request.[11] It's speculated that others throughout history used this practice to build immunity to poisons including Visha Kanya (maidens in India who were used as assassins) and Russian mystic Rasputin who survived a poisoning attempt.

Gretel continues to grow in her bravery throughout *Gretel and Hansel* and takes more risks in her waking and her dreaming. What is the science behind bravery versus fear? According to Mara Myers, "courage is what's required when the thing you fear is emotional and

Ricin, brought into the zeitgeist by its use on the show *Breaking Bad* (2008–2013), is one of the strongest poisons in the world. A twenty-year-old man was found to have an inherited metabolic defect protecting him from the deadly compound.[12]

there is no real physical danger. Bravery is what's required when the threat is real and poses a physical danger. In the end, bravery requires the repression of fear, but it takes courage to explore the fear you had to repress in order to be brave."[13] Can we actually make ourselves braver? Yes! Neuroscience proves that courage is a cognitive ability that can be strengthened or atrophied. All we need is a willingness to act and the ability to manage our anxieties regarding a situation. Thankfully, Gretel was able to overcome her fear in order to defeat the witch and save herself and her brother from their demise.

CHAPTER NINE
MALEFICENT

Maleficent has been a villainess by various names in stories beginning in the early seventeenth century. She has been portrayed as hysterical, vengeful, vain, and murderous. In every retelling, Maleficent was pitted against the opposite archetype: a beautiful, young, obedient woman. As the author of *The Secret History of Maleficent: Murder, Rape, and Woman-Hating in Sleeping Beauty* said:

> We are taught to revere youth, beauty, and passivity among women, and to be suspicious—hateful even—of women who are older, powerful, and opinionated. Women who stand up for themselves are equated with murderous child-killers, witches, and ogres, while women who literally lay back and take what men give them are rewarded for their "virtue."[1]

Audiences may recognize Maleficent from Disney's *Sleeping Beauty* (1959) most famously when she cursed the title character by saying "The princess shall indeed grow in grace and beauty, beloved by all who know her. But, before the sun sets on her sixteenth birthday, she shall prick her finger on the spindle of a spinning wheel and die!"

Just like *Wicked* (2003) allowed audiences to gain empathy for the Wicked Witch of the West, the films *Maleficent* (2014) and *Maleficent: Mistress of Evil* (2019) allow us to see the backstory of this previously one-stroke character. Written by Linda Woolverton, *Maleficent* explores the rich and complicated backstory of the iconic villain and gives her depth and nuance. About playing the title character, Angelina Jolie said, "Part of the thing with this role is that you realize that there's no halfway. If you're going to do it, you can't kinda do it. You have to just go fully into it and enjoy it. The original was done so well, and her voice was so great, and the way she was animated was so perfect that,

A study on comas showed that the longer a patient remains in a coma the poorer his or her chance of recovery and the greater the chance that he or she will enter a vegetative state. By the third day the chance of making a moderate or good recovery is reduced to only 7 percent, and by the fourteenth day is as low as 2 percent. By the end of the first week almost half of those patients who have not recovered consciousness are in a vegetative state.[2]

if anything, I just was worried that I'd fail the original." When Aurora (Elle Fanning) meets Maleficent for the first time, she assumes she is her fairy godmother. What are the cultural views of fairy godmothers or guardian angels? Fairy godmothers often assume the duties of protector

"The earliest recorded mention of fairies comes from 1000 BC in The Iliad, where Greek poet Homer wrote 'watery fairies dance in mazy rings.' Many creatures that appeared in ancient Greek myths, such as satyrs, nymphs, and sileni were also considered to be fairies, however."[3]

and guide. Angels appear in the Bible and carry out duties on God's behalf including executions, delivering messages, and acting as guards. Other cultures have similar stories such as the fravashis of Zoroastrianism, the karabu of the Assyrians, the Daimones of the Greeks, and the genii of the Romans. An Associated Press poll in 2011 revealed that 77 percent of adults in the United States believe in angels and sociologists think they know why: "Angels pervade popular culture in books, television shows, and movies . . . Believers exchange informal testimonials in newsletters and interpersonal conversations about the potential power of angels to influence the world, and more than half of Americans (53 percent) believe that they have personally been saved from harm by a guardian angel."[4] It is easier to believe in something if we are constantly surrounded by reminders of it.

Maleficent grows to be a mother to Aurora throughout *Maleficent* and tries to take the curse back but cannot. It's an empathy building sequence that humanizes her and makes the audience think about guilt. What is the science behind it? Our emotional responses are governed by the limbic system in our brains. This area of the brain is short-term minded, like a Post-It Note, and it's where the feelings of guilt thrive. It's less rational and can run rampant if not checked. Three ways to handle feeling guilty according to science writer Colin Robertson are: be kind to yourself instead of beating yourself up, learn from your mistakes, and forgive yourself.[5] Neurologists agree that human emotions have

The makeup for *Maleficent* took about three hours to apply and one hour to remove every day. Angelina Jolie was in full makeup for seventy days, often for sixteen hours at a time.[6]

evolutionary functions that have served us toward survival so it's important to pay attention to them and learn from them. Maleficent definitely paid attention to her own feelings toward Aurora and acted accordingly.

When the curse takes hold of Aurora in the film, Maleficent can sense it. How can people have a sixth sense about certain things? Neuroscientist Eric Haseltine believes we all have more intuition than we realize. We can sense things about people by unconsciously reading their nonverbal communication through scent and through sound. We get a "gut feeling" about a situation and he encourages us to trust it: "It's useful to be aware that we are aware of things—both learned and innate—even though we aren't aware of *why* we are aware, because such awareness can decrease harmful self-doubts . . . The bottom line is that there are scientifically valid reasons to *trust* your feelings, perceptions, and intuitions, even when you can't sense how you sense them."[7] There are anecdotal stories of people being able to sense when something has happened to a loved one, but no scientific proof exists yet to explain this phenomenon.

An incredible rewrite of the original story involves "true love's kiss." It's the only thing that can awaken Aurora in *Maleficent*, but it's not a kiss from Prince Philip (Brenton Thwaites). Maleficent gives Aurora a kiss on the forehead, after telling her how much she will miss her smile, and Aurora awakens. "No truer love . . ." says Diaval (Sam Riley) as tears stream down all our faces. (I'm not crying, you're crying!)

Maleficent is invited in the second film *Maleficent: Mistress of Evil* to meet now engaged Aurora's future in-laws for a meal. She needs to use personality mirroring when she meets the king and queen for the first time. Why do some people do this? When we communicate with others, sometimes we take on their affectations or mannerisms. It's a natural way to build rapport that we may not even be aware that we're taking part in. For others, it's a strategic method to show others you are like them and is sometimes used by salespeople to improve business. Mirroring affects our brains and fosters feelings of closeness and trust but if it's fake or overly obvious it can create the opposite effect.

As a powerful fairy, Maleficent is allergic to iron. Is this possible in humans? People do have allergies to various metals, especially nickel, cobalt, and chromates, and will typically get a rash from contact with one. Iron contact allergies are less likely but appear throughout folklore as a way to trap or repel fairies, witches, ghosts, and other creatures. If you ever meet someone with this rare allergy you can ask them which of these creatures they are! Maybe not. Use personality mirroring first to determine if they are on the same page about science and horror facts.

The dark fey in *Maleficent: Mistress of Evil* deliberately drop their prey to kill it. Many species of birds are known to do this and open hard-shelled prey by dropping the prey repeatedly onto the ground from considerable heights.[8]

CHAPTER TEN
THE WRETCHED

Fairy tales involving witches have pervaded children's bedtime stories for centuries. Witches have been portrayed as misunderstood, downright evil, or somewhere in between. With their new take on witches, Brett Pierce and Drew Pierce created *The Wretched* (2019). The audience follows a skin-walking witch over time to learn about her unique powers and fixations. We had the opportunity to interview the writers and directors of *The Wretched*, the Pierce brothers, about making this movie.

Kelly: **"We absolutely love your movie so thank you for taking the time to talk to us! I was wondering if you could each tell us about your first memory of being introduced to witches?"**
Drew Pierce: "For me it was Roald Dahl's *The Witches* (1983). I was just obsessed with that book, and I grew up dyslexic. I didn't like to read all that much, but I read that book, I don't know, five or six times when I was a kid. It was so spooky, and it felt like something taboo, like I wasn't supposed to be reading it. It felt a little scarier than a lot of the stuff I was reading when I was seven.
Brett Pierce: "The one that I always think of is Meg Mucklebones from *Legend* (1985). That portion of that movie just always stuck with me in my imagination my whole life. I think that's honestly why we wanted to make *The Wretched*. I wanted it to be a creature representation of a witch because that was the iconic witch to me."

Meg: **"I always want to know when people are writing together, how do you guys do it? What is your process like?"**
Drew Pierce: "There's a lot of banter back and forth, pitching ideas all the time. But that's just sort of like brothers. We're always throwing ideas out there like 'would it be cool if we made a movie like this?' Typically, the way we do it, he [Brett] usually writes the first draft. I'm the crazy

outline preparation guy and he's like 'dive right in! I don't even want to think about it, I just want to start typing.'"

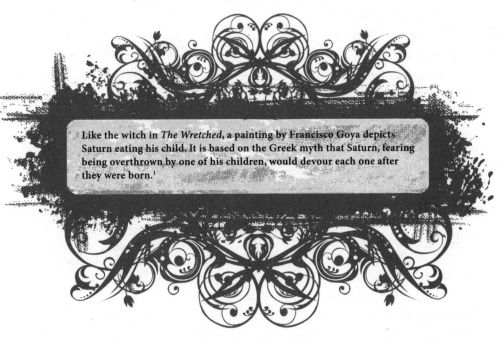

Like the witch in *The Wretched*, a painting by Francisco Goya depicts Saturn eating his child. It is based on the Greek myth that Saturn, fearing being overthrown by one of his children, would devour each one after they were born.[1]

Brett Pierce: "It gets boring to me if I know exactly what's going to happen everywhere."

Drew Pierce: "It's funny, we have such polar opposite approaches, but I think it works for us in a way. He usually dives in and sometimes I mean he's written entire scripts . . . and then I rip out like five great ideas . . . and completely throw out the rest of it and then I restructure."

Brett Pierce: "Then I get mad at him!"

Drew Pierce: "Then I restructure it from scratch, do a whole plot, then a lot of times we'll kind of write back and forth. He'll write for five, ten pages and then I'll write for five or ten pages. It depends on who has the best idea for the scene. Or there's a lot of times where one of us will be like 'I just don't think this works' and the other one's like 'just let me take a shot at that, I know it works.'"

Brett Pierce: "So it's a lot of bad first drafts by me!"

Kelly: **"That's very cool to hear! We have similar styles. How much research did you do on actual witchcraft before writing the movie or was it all imagination?"**

Drew Pierce: "I've had a long fascination with a lot of folklore. It's something I like to draw from because it's got this interesting aspect. We live in a world of technology so there's not a lot of unknowns anymore or creepy places in the world. But if it's folklore, and it's about the past, you kind of get this element of scary because nobody understands it completely. So, it was me reading a lot about various witch myths from around the world like the Cailleach, Black Annis, Baba Yaga, and Jenny Greenteeth. I wanted to make a story about one of these witches because they all have such

The trope of a witch riding a broomstick has roots in science, specifically broomsticks purportedly being used to administer hallucinogenic salves that caused the user to fall asleep and dream of flying.[2]

unique elements about them or rules about how they operate. I felt like people weren't really grabbing from the actual folklore tales about the witch."

Brett Pierce: "Mostly for us she's a Black Annis because she lives in this bower of a tree, and she steals children and eats them. Drew also found this witch called the boo hag which we were just super fascinated by. She was basically like a skin-walking witch. She would wear somebody else's skin, and you know the old myth is like this husband discovers that his wife might not be who she says she is and she's hanging up her skin in the closet so he salts it one night and when she comes back and crawls into her skin it burns her . . . We took that and the idea of a skin walker because we liked it and thought there's two rules that we can use that nobody else had been using, at least recently, for witches. We were really fascinated by that.

Then, making people forget about their children was kind of born out of a plot problem. We were just kind of like 'well if somebody's stealing and eating all these kids, how does nobody catch on to this?' It's like we couldn't figure out how nobody's noticing that all the kids are disappearing, and Drew was just like 'oh, well she makes them forget! That's one of her powers!'"

Ty (Kevin Bigley) begins bleeding out of his ear after his wife Abbie (Zarah Mahler) whispers a spell into it in *The Wretched*. In Irish legends, a banshee, a female spirit who heralds death with her scream, can cause people's ears to bleed.

Kelly: **"One of the things that impressed us the most about this movie was your practical effects. We were wondering about the process from your imagination to reality. How did you do this? It was amazing!"**

Brett Pierce: "It was just dark magic! No, Drew and I grew up around special effects as kids. Our dad was the special effects artist for the original *Evil Dead* (1981), so we were always really big fans of practical effects. We always thought they worked better for horror movies. They kept the realism of making you feel like this is really happening.

Drew has a really good character design background. He's worked in animation for years and years so he initially did a bunch of creature designs of what we wanted her to look like. Then, we met this really fantastic special effects artist named Erik Porn who worked on *Teen Wolf* (2011–2016) and *The Walking Dead* (2010–) and we approached him and talked to him. He's a fantastic sculptor and he would just start sculpting the creature based on Drew's designs. It was great because we were back in Michigan prepping to shoot the movie and Erik was still working on the molds in LA. He would send us pictures and Drew would literally bust out his Cintiq [creative pen display] and draw over the sculpts as they would go. He was very aware of how we didn't want creature effects like in *Star Trek* (1966–1969). Sometimes there was amazing stuff back in the day but sometimes it gets bulky and starts to feel fake like there's a bunch of stuff on this person's face or around their body. It doesn't feel natural. We really tried to build to the body of the performer who was underneath, not go too heavy, so it felt like a real person with real dimensions. It was a mix of constantly being in contact with him and constantly changing the design.

"Drew's a very accomplished storyboard artist for other movies so we did what every practical effects artist wants and never gets: we storyboarded every effect shot and said this is what we're gonna show. We won't show other angles. It was all super challenging because it took hours and hours. Every time it took like four or five hours to get her into that makeup. The effects scenes, whenever you shoot them, they just take longer no matter what."

Drew Pierce: "It's the most exciting part of a horror movie because everybody shows up when there's a special effect. Everybody's amped and everybody gets really immature when you're like 'we're about to kill somebody' and they're like 'oh, can I come watch?'

Meg: **"We love that! I'm curious why you chose to film in Michigan and what that meant for the storytelling. I thought it was a great juxtaposition of this beautiful, on-the-water look, and then all these evil things are happening."**

Drew Pierce: "Yeah, one of the things we talked about before going into this movie is that we didn't want to start the movie full of dread and just creep like a lot of movies, you know? They show up at the haunted house and you're like 'who would stay here?'"

Brett Pierce: "Yeah, and it's like dripping. You look at the house and you're like 'that's got bodies in the wall why would you go here, everything's dark and scary?'"

Drew Pierce: "When you're making an indie movie, you want it to feel like it has a little bit of scope and some different cool locations. One of the ways we knew we could pull that off is going back to Michigan because that's where we're from. We shot where my in-laws run the local sailing school for young kids and it's a great location for this little creepy story."

The legend of Jenny Greenteeth involves a river hag with green skin that would pull children into the water and drown them.[3]

Kelly: **"Your movie has** *Rear Window* **(1954) vibes. What other movies were you pulling from for inspiration?"**

Drew Pierce: "I think the two most obvious are like *Rear Window* and *Fright Night* (1985) just because it's the voyeuristic person next door. That works so well for horror because you can slowly immerse people in this creepy world."

Brett Pierce: "We learned things along the way, too. The challenges that arise in a POV movie like *Rear Window* or *Fright Night*, you realize they built their sets to do that. They built bigger windows and all these other things so you can look through stuff. Lots of weird challenges come up in an indie movie. You're like 'how do I find two houses that are next to each other in the country or in the middle of nowhere that have big enough windows that are looking at each other?'"

Meg: **"What are your future or current projects?"**

Brett Pierce: "We've been working on a lot of different stuff. I mean it's the honest truth with indie guys, we're trying to make the transition to hopefully get to make a studio movie. The movie did open up some opportunities but it's still that hard trick. The things we're working on right now are *Wretched Two* and we've been working on a werewolf thing."

Thank you to the Pierce brothers for the informative and fun interview!

Two of the inspirations for the film, the Black Annis and the Cailleach, sparked our need to know more about the legends. The Black Annis is from English folklore and is a witch that has a taste for human flesh. She has iron hooves, steals children away during the night, and wears their flesh around her waist. The legend was used as a cautionary tale to try to get children to behave because if they were naughty, they'd get snatched away. The legend of the Cailleach comes from Gaelic mythology and translates to "old woman" or "hag." She is associated with the winter, herds deer, and fights spring. According to folklorists:

> She continued to function as a transgressive alternative to the later dominant Christian culture, providing a therapeutic function for the community through story and symbolism, as well as supplying models for female agency through its derivatives of the *bean feasa* (wise woman), *bean ghlúine* (midwife) and the *bean chaointe* (keening woman). Though now absent in cosmological considerations, the figure continued in local folklore, especially in tales which served as explanations of geological forms and placenames, but also in the 'When all else fails,' wise woman stories.[4]

Salt is used to ward off evil in this, and several, horror movies. What is the history behind this process? Salt has always held value throughout history. Easy access to it during meals was a sign of status in medieval times and was viewed in many religions as representing purity. The Bible mentions salt in Leviticus 2:13: "And every offering of your grain offering you shall season with salt; you shall not allow the Salt of the Covenant of your God to be lacking from your grain offering. With all your offerings you shall offer salt." Salt is used to create magic circles in witchcraft that offer protection, contain energy, and form a sacred space. In 2019, a man in Florida took this belief a little too far. He was arrested after pouring salt on his feet and all over the floor of a Walmart to "ward off evil spirits."[5] He retreated to lay in the woods before his arrest. Hopefully it worked for him!

SECTION FOUR
WITCH AS MONSTER

CHAPTER ELEVEN

THE AUTOPSY OF JANE DOE

At the very cusp of the climax of *The Autopsy of Jane Doe* (2016) coroner Tommy Tilden (Brian Cox) and his son Austin Tilden (Emile Hirsch) realize what the markings are on an old cloth swallowed by a deceased woman, Jane Doe (Olwen Kelly). Written in Roman numerals is a reference to the Bible's Leviticus 20:27: "Any man or woman who makes use of spirits, or who is a wonder-worker, is to be put to death: they are to be stoned with stones: their blood will be on them." Tommy reads the passage aloud and they realize their worst fears, that this seemingly dead woman, a scientific impossibility in so many ways, is, indeed, a witch.

Before the supernatural events unfurl, leaving the main characters dead, Jane Doe, lacking any clothes or identification, is brought to the Tilden Funeral Home by the local sheriff (Michael McElhatton). No one

The first use of "John and Jane Doe" to name those who were unidentifiable was in British law, in which a process known as "action of ejectment" made it simpler for landowners to file legal complaints against a fictitious squatter rather than a real person. There is no known reason for the choice of name.[1]

understands why she was found in the dirt basement where a murder occurred, so Tommy and Austin are tasked with staying up all night to find evidence of her cause of death.

While outwardly Jane Doe appears perfect, pale but unmarked skin, nary a scratch or bruise on her, it becomes clear to the men of science that something strange is afoot. Her wrists and ankles are fractured, though there is no swelling or bruising on the skin. She also has lacerations in her vagina, which leads them to wonder if she was a victim of human trafficking. As they continue their four-part examination, there are inconsistencies that worry them. She has an unusually small waist, which they attribute to a corset, though those have long gone out of fashion, since about 1910 in fact. Her lungs are scorched and blackened, which Tommy notes is too severe to come from cigarette smoking. Also, her clouding eyes suggest she was dead several days, yet her lack of rigor mortis is confounding.

Fully developed rigor mortis is an easily identifiable and reliable indicator that death has occurred. The time of onset is variable, but it is usually considered to appear between 1 and 6 hours (average 2–4 hours) after death. Depending on the circumstances, rigor mortis may last for a few hours to several days. The muscles of the face and neck are often the first to be affected and the rigidity spreads backward over the trunk and limbs. Relaxation of the muscles occurs in roughly the same order.[2]

When Tommy cuts into Jane's chest to make a Y-incision, she bleeds. This shocks Austin because livor mortis, another standard indicator of death, should have made it impossible for blood to flow:

Livor mortis is technically the purplish-blue discoloration of the skin in the dependent parts of the body due to the collection of blood in skin vessels caused by gravitational pull. Hypostasis develops as spots of discoloration within half an hour to 2 hours. These spots then coalesce to form larger patches, which further combine to form a uniform discoloration of the body's dependent parts that have not been subject to pressure, which appears from

6 to 12 hours. The discoloration becomes "fixed" after a certain period, owing to the disintegration of blood cells and the seepage of hemoglobin.[3]

Giles Corey is the only known American to die of pressing, but it was used in Britain in the sixteenth century, as well as in Asia in the nineteenth century, in which elephants were used to crush those convicted.[4]

Along with the discovery of the Bible verse near the end of the film, perhaps the most frightening indication of Jane Doe's state of being is when Austin and Tommy cut off a piece of her brain. Beneath the microscope they see that her brain cells are active and moving, which is proof that despite her body being removed of organs, she is alive. Somehow, she is feeling every infliction of pain, hundreds of years after her capture.

Through the clues established, they come to understand that Jane Doe was tortured; stabbed, made to ingest poison, and burned, among other indignities. Tommy wonders if she was innocent but formed into a vengeful witch by the acts of her torturers. He ponders this tragic irony before Jane makes him feel every pain she has suffered. This is described by film reviewer Mara Bachman in an article for *Screenrant*:

The body of Jane Doe lives on to enact revenge on any and all people who attempt to take away her bodily autonomy, whether through an autopsy or otherwise. She stands as a symbol that works against the brutality of the Salem Witch Trials. *The Autopsy of*

Jane Doe offers unique insight into the brutal workings of one of America's oldest supernatural histories while providing a modern twist on the lore surrounding witchcraft and one of the United States' darker historical eras.[5]

The Salem Witch Trials have come up often in our research as its history is indelibly linked with our Western idea of the witch. Twenty-five deaths were attributed to the trials, nineteen of which were hanged, five perished in jail, and one was crushed to death. There is no proof of someone enduring the torture of Jane Doe, but that does not make it unlikely. While there is a supernatural element to Jane's destruction, her physical pain could easily be a metaphor for the emotional turmoil and untimely deaths of innocent people. And more literally it could reflect the horrific death of Giles Corey. Corey, like his wife, Martha, was accused of witchcraft. He was eighty years old and perhaps because of his age chose to be obstinate in court. He remained mute, refusing to answer the judge. This led to him being sentenced to "peine forte et dure" which is being pressed to death with large stones. It took two days of pain and humiliation before Giles Corey succumbed to his injuries in the field beside Salem's jail. It is said that his final words were "more weight."[6]

This archaic way to die must have been endlessly painful and certainly reflects the torture Jane Doe withstood. If anyone should come back and do some haunting, it is Giles Corey!

As I (Meg) watched the movie unfold, I became fascinated with the process of the autopsy. At its root it is a scientific process, using chemistry, physics, and biology to make sense of death. Yet, there is a Holmesian aspect to its mysteries that, even in the most average of circumstances, can be as imprecise and fascinating as crime investigation. It is Sherlock Holmes and Dr. Watson in one, a vital

Jimsonweed, also known as Devil's Snare, is found in Jane Doe's stomach. This poisonous flower was brought to America from Britain and has long been the culprit of many deaths, including a group of unknowing, hungry soldiers in 1676, who ate the flower and its spiky fruit full of deadly alkaloids.[7]

profession. The concept of dissecting the deceased goes back centuries. "Greek physicians Erasistratus and Herophilus pioneered anatomical autopsies by dissecting cadavers to study the workings of their organs and nerves. Later, second century physician Galen became the first to link patients' symptoms with autopsy observations."[8] Through his examination of the dead, Ancient Greek Galen of Pergamum developed the notion of human fluids, blood, phlegm, yellow bile, and black bile, being at the center of all diseases. This was not disputed for fourteen hundred years, as during the Middle Ages the concept of autopsy was considered uncouth. Bodies were to be respected and buried without an examination. It wasn't until the 1800s that the modern concept of autopsy flourished. Doctors and scientists began to understand how the body told a story.

> Introduction to the modern-day autopsy was pioneered by Karl Rokitansky of Vienna, who had completed more than 30,000 autopsies and supervised about 70,000 autopsies during his career. Rokitansky was the first to examine every part of the body, with a systematic and thorough approach. However, it was his competitor, Rudolf Ludwig Karl Virchow, who used microscopy to examine each organ carefully. This application allowed Virchow to show evidence that cellular pathology was the basis to the understanding of disease.[9]

With the advent of DNA as well as other advances, the autopsy has become a powerful tool that can solve murders and medical mysteries. One example is the long-ago death of an Egyptian mummy. In 1825, Dr. Augustus Bozzi Granville made a splash in the scientific community when he conducted the first autopsy on an ancient mummy. He was able to determine that Irtyersenu or "lady of the house" was in her fifties when she died. He concluded that a mass in her ovaries was the likely cause. This made headlines around the world, but in the 1990s, Dr. Helen Donoghue discussed her own autopsy of the same mummy with BBC News:

> Today's tools of genetic analysis have enabled her to probe deeper than Granville was able to. Donoghue extracted DNA fragments

from the mummy, which proved to be those of the bacterium that causes tuberculosis (TB). TB is a highly contagious illness that mainly affects the lungs. The TB DNA had spread throughout the Egyptian woman's body. Such an infection would have been fatal.[10]

The ovarian mass was found to be benign. These two findings show how exponentially the science of autopsy has grown.

One aspect of *The Autopsy of Jane Doe* that I found impressive was the performance by Olwen Kelly. As Jane Doe, she was tasked with the difficult job of lying on an autopsy table in the nude, pretending to be dead, for the majority of the film. The film's director, Andre Ovredal, expressed his own surprise at Kelly's dedication:

> She had to lie there on a slab for maybe eight, ten hours a day for weeks on end. And it was a cold marble table. She didn't say a word, she just lay there, and was so amazing with everybody, and she made everybody calm. Because it's kind of an awkward situation, you know, with the nudity, and everything. Her personality made everybody relax.[11]

Apparently, Olwen Kelly is a yogi and used her knowledge of shallow breathing to appear dead! In an article for *SyFy Wire*, Bryan Reesman explores how important it is that actors really appear dead:

> A good dead performance is something audiences take for granted, but a dicey one can pull them out of the realism of the scene. A great example is a circular pan shot in the entertaining 1972 film *Raw Meat* (starring Donald Pleasance with an appearance by Christopher Lee) about a cannibal roaming below the London Underground. A deceased man, eyes open, sits against a wall as the camera does a full 360-degree pan around the flesh eater's lair. When the rotating shot finally returns to the victim, he blatantly blinks.[12]

This kind of movement instantly takes the viewer out of the world constructed by the filmmakers and can ruin a decent shot. (I know that a not-so-dead actor blinking or breathing has bothered me many times!) Reesman goes on to interview actor Matthan Harris who has portrayed a corpse on screen in several films like *The Inflicted* (2012). Harris is quick to say that Olwen Kelly should win an award for her performance.

Dummies are often used in death scenes, particularly when the victim goes through extreme trauma (think a long fall or an exploding head). Yet, many directors, like Andre Ovredal, insist on the realism of a human. Lying still may seem like an easy job, but controlling the uncontrollable—blinking, breathing, flinching—is next to impossible. Thankfully, Olwen Kelly did splendidly in *The Autopsy of Jane Doe*, which keeps the realism, and therefore scares, on track.

On its face, *The Autopsy of Jane Doe* can be enjoyed as a supernatural thriller. There is the walking dead in the Tilden morgue navigating through the dark, and even an accidental axe murder when Austin's girlfriend Emma (Ophelia Lovibond) is mistaken for a ghoul. The witch is presented as a monster. She kills the main characters, not just taking their lives, but relinquishing her own anguish and pain onto them. Yet, even Tommy who endures the most, sees the perversion of justice that has occurred. Jane Doe was made into a beast, much like Frankenstein's

Seventeenth-century Italian physician and autopsist Antonio Valsalva, lacking chemical tests and the understanding of germs, sometimes tasted the fluids he encountered in cadavers in an effort to better characterize them.[13]

monster, and the puritan society in which she lived was the maniacal doctor. Through the extreme torture she endured, Jane was given her witch powers. To make others feel her indignities is her only way to avenge what happened in Salem. At the end of the film, we see the slightest tick of movement in her toe, suggesting she is growing stronger, and perhaps, will rise from the morgue's table and truly inhabit her calling as a witch.

We can only hope.

CHAPTER TWELVE
DON'T KNOCK TWICE

In the British supernatural horror film *Don't Knock Twice* (2016), it isn't long before the characters do just that. "Knock once to raise her from her bed, knock twice to raise her from the dead." This rhyme is uttered outside the door of a long-dead woman who took her own life, and was believed to be a child-eating witch.

Unfortunately for troubled teen Chloe (Lucy Boynton), this legend proves to be true, and by knocking on the door she has become prey for the witch living within. And this is no ordinary fairy-tale witch with an oven ready for plump children. It is the Baba Yaga of Eastern European myth, gnarled and monstrous. She crawls rather than walks, her naked body more spider-like than human. Her goal is to devour the innocent, especially those who dare knock upon her door.

Who is this Baba Yaga, and how has she endured from ancient Slavic folklore to modern film? The origins of her name are just as varied as the stories told. "Baba" is believed to be an antiquated word for "old woman" while "Yaga" has little linguistic relevancy. It could be a word for "snake," though experts have not agreed on a single meaning. The same could be said for the witch herself, as she is often depicted as a cannibalistic crone, while other times in folklore she shows mercy.

In the tale of Vasilisa the Beautiful, arguably the most famous story in which Baba Yaga appears, Baba Yaga takes on several, seemingly conflicting roles. Beautiful Vasilisa lives with her wicked stepmother and two homely stepsisters, who all conspire to have her killed. After several unsuccessful attempts, they finally send Vasilisa directly to Baba Yaga's hut, knowing that the crone eats humans "as one eats chickens." But instead of devouring the girl, Baba Yaga forces her to do a series of seemingly impossible menial tasks, such as separating grains of rice from wheat kernels before dawn. When Vasilisa succeeds at this, she's granted one of the skull lanterns that rings Baba's house; upon returning home, the lantern immediately engulfs her horrible family in flames, freeing her from their tyranny. Eventually, beautiful Vasilisa ends up marrying the Tsar.[1]

Baba Yaga has existed in many iterations throughout not only ancient myth and folklore, but within modern media. In the comic book series, and later in the 2019 film *Hellboy*, the witch known as Baba Yaga (played by Troy James and voiced by Emma Tate) has a proverbial bone to pick with Hellboy (David Harbour), as he previously caused her to lose an eye. Now, one-eyed and ugly as ever, she is working to help his enemies. In an article for *Entertainment Weekly*, the creator of the comic book series, Mike Mignola, explained that he had been fascinated with Baba Yaga since a fellow student in fifth grade did a report on the fearsome witch. Journalist Christian Holub also spoke with the film's creature design and special makeup artist Joel Harlow, who was committed to making Baba Yaga as scary as possible. "She's a Russian witch and I knew I didn't want to go with a stereotypical witch-looking character with a hook nose and pointed chin. I wanted to go with something much more frightening and much creepier, something that really unsettled me."[2] The result of Harlow's work is an effectively terrifying witch with ghoulish features, pallid skin, and stringy hair.

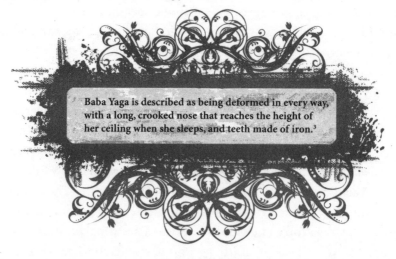

Baba Yaga is described as being deformed in every way, with a long, crooked nose that reaches the height of her ceiling when she sleeps, and teeth made of iron.[3]

In the science fiction series *The Magicians* (2015–2020), Baba Yaga rejects any sort of cumbersome body of an ancient hag, and instead, uses her supernatural power to inhabit the young body of Bailey (Olivia Norman). Possessed by Baba Yaga, Bailey hunts for artifacts to appease her, which the witch greedily takes.

For their 2017 article "Identifying Impressions of Baba Yaga: Navigating the Uses of Attachment and Wonder on Soviet and American Television," researchers compared how Baba Yaga is perceived in both Russia, where she originated, and in America, where the witch has appeared in many forms. One vital piece of their research was to understand the difference between how Soviet Russia and America consumed TV:

> The function of the state is distinctive in the Soviet examples, and Soviet televised narratives of Baba Yaga served propagandistic purposes. Although the narratives were for entertainment, they also taught socialist principles suited to the Soviet Union. The protagonists, just like those in the traditional tales, outwit Baba Yaga through problem solving, ingenuity, courage, and hard work. These traits were valued in Soviet citizens. Baba Yaga on Soviet television allowed producers to show how these socialist principles of hard work, innovation, and aptitude based in peasant culture could be applied to any time period, including their own. For example, televised adaptations of Baba Yaga narratives also included revisions in the way the protagonist escaped from Baba Yaga, exhibiting a modern Soviet influence on the televised narratives, and allowing viewers to identify more with both the protagonist and Baba Yaga.[4]

The researchers recorded nostalgic comments about these Russian cartoons on YouTube, coming to find that those who watched these depictions of Baba Yaga growing up were not enchanted by the propaganda, but by the witch's ambiguous and fascinating character. In contrast, American cartoons treat Baba Yaga as an "other," as she is foreign and deeply rooted in Russian culture.

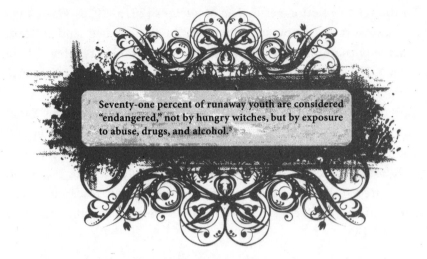

Seventy-one percent of runaway youth are considered "endangered," not by hungry witches, but by exposure to abuse, drugs, and alcohol.[5]

While Baba Yaga has been introduced to Russian and American children in varying ways, the article contends that both countries, however different, have embraced the magically dark Baba Yaga and that she will only continue to seep into the modern cultural zeitgeist.

In *Don't Knock Twice*, Baba Yaga isn't distracted by baubles or revenge, she is only interested in feasting on children. Her powers are vast, as she can open the thresholds between worlds, represented by any simple door. She can also mark a human with a symbol in order to do her bidding. They are considered her "slave" who can only be freed by her magic if they kill themselves or trick another to take their place.

When Tira (Pooneh Hajimohommadi) gives Chloe's mom, Jess (Katee Sackhoff), a "talas" charm to wear to improve her relationship with her daughter, it is not revealed until later that this charm is the mark of Baba Yaga. Shaped like an evil eye, it burns into Jess's flesh, transforming her into the witch's human slave.

The mark of the evil eye goes back centuries. Its oldest known iteration is a charm with an ocular design from 3330 BCE in Mesopotamia (modern-day Syria). The symbol has been used in a number of cultures to ward off evil. The Egyptians would bury their pharaohs with the "eye of Horus" for safe passage to the other world. Phoenicians wore blue beads decorated with eyes as spiritual protection. "We're still affixing the evil eye to the sides of our planes in the same way that the Egyptians and Etruscans painted the eye on the prows of their ships to ensure safe passage. It's still a tradition in Turkey to bring an evil eye token to newborn babies, echoing the belief that young children are often the most susceptible to the curse."[6]

So, what is this curse? The evil eye represents envy. As a person grows successful, rich, and happy, they are looked upon by their friends and neighbors with growing jealousy. If they do not protect themselves with an amulet of an eye, they will lose all their fortune and good will. Unfortunately for Jess and Chloe, the charm is not a protection, but rather a mark of who will be controlled by the malevolent witch.

As serial killers prey on those inhabiting the fringes of society, it makes a sort of macabre sense that Baba Yaga would feast on children who may not be missed. Police detective Boardman (Nick Moran) even implies that ten-year-old Michael (Callum Griffiths) probably ran away rather than be the victim of any supernatural force. He was, after all, one of the "forgotten," with no family to care for him.

To be considered a "runaway," a child under the age of fourteen must stay away from their home for one night without contact, or for teens aged fifteen and older, two nights. While this sounds like a simple enough categorization, there are many factors that come into play when grouping statistics of runaway children. A large-scale American study was done in the late nineties, in which demographics, as well as prior abuse, housing, and repeated episodes were studied. Interestingly, only 18 percent of runaway incidents were children under fourteen, the vast majority being aged fifteen to seventeen. Also, most times the children ran away was brief, they stayed close in proximity to their home, and police were not involved. Only less than 1 percent were never found, a startling statistic when you think of the media's prolific portrayal of abducted children. Why do children run away from home? In a more recent study from the National Runaway Safeline, the evidence is clear: 47 percent of children experience conflicts with parents or guardians at home, 34 percent of runaways experience sexual abuse at home (80 percent of those girls), and 43 percent of teens report physical abuse as one of the main reasons they left home.[7]

Abuse is clearly at the heart of why children end up in dangerous situations, sometimes choosing homelessness over living with those who abuse them. It's no wonder that Jess is suspicious when Detective

Boardman says ten-year-old Michael ran away, as after a decade missing, this would be highly unusual for a child his age. *Almost* as unusual as a slobbering, deformed witch keeping him locked in a cage until she's ready for dinner.

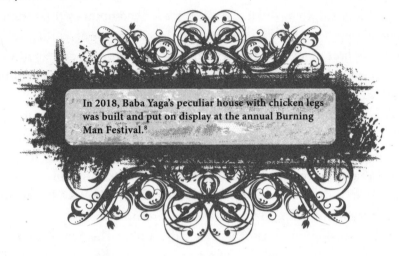

In 2018, Baba Yaga's peculiar house with chicken legs was built and put on display at the annual Burning Man Festival.[8]

The theme of abandonment and strained motherhood prevails, as Jess is far from the ideal mother. She does not literally devour her child, but she has harmed Chloe with her selfish addiction and abandonment. Baba Yaga is often linked with motherhood, as she is as ambiguous in her love as Jess is. In the story of Vasilisa, Baba Yaga is strict, but caring and maternal. In other legends, she is the witch of nightmares. This dichotomy is what makes her frightening. It is this same confusion that Chloe lives with, remembering the good times with her mother, only to be struck with the anguish of the neglect. As the title suggests, *Don't Knock Twice* deals heavily with door imagery. With magic,

There are many mythical entities in charge of thresholds, like Roman god Janus, who was believed to protect every gate and door in Ancient Rome.[9]

Baba Yaga enters doors from her own underground, hellish lair. In legend, she would be living in a house atop chicken legs, but for the purposes of a modern horror film, her domain is scarier and less whimsical. In fact,

Jess and Chloe even burn all the doors in their house to avoid Baba Yaga's control. Doors and openings have long held metaphorical significance, from the frightening gates of hell to those ivory pillars of heaven and everything in between. They often represent a new beginning or ending, and in folklore are synonymous with moving between worlds. In the film, Baba Yaga uses this power to transport Jess to her lair, effectively smuggling her out of prison through supernatural means.

The ending of *Don't Knock Twice* is effectively bleak. Like Baba Yaga's own ambiguity, it ends with many questions never answered. The witch, in all her horrendous glory, is the final one to knock. This Eastern hag is more formidable than any Wicked Witch of the West, her looming presence forever overshadowing her human prey.

CHAPTER THIRTEEN
THE CONJURING

As the title of the film suggests, *The Conjuring* (2013) is a witch's tale at its dark, terrifying heart. Jump scares and creepy dolls aside, it is the story of a witch bent on destruction.

The beginning of a horror franchise, *The Conjuring* not only scared our collective pants off, but introduced many viewers to the real-life inspirations Ed and Lorraine Warren (played by Patrick Wilson and Vera Farmiga). The couple devoted their lives to paranormal research and founded the New England Paranormal Research Headquarters in 1952. There is a brief biography on the headquarters's website about Ed and Lorraine that describes their earnest appeal:

> Ed Warren was a demonologist. Lorraine Warren was a trance medium. They were not occultists. They were not strange. If you had the privilege of speaking to them, they would seem like normal folks with regular jobs. They were ordinary people who happen to do extraordinary work in a field that most people fear or don't believe. The forces they confronted are religious entities that—by their own admission—exist for the sheer purpose of opposing the works of God.[1]

This dichotomy of down-home charm and a passion to fight against demons is perhaps why the Warrens became such popular guests on shows like ABC's *Scariest Places on Earth* (2000–2006) and Discovery's *A Haunting* (2002–), as well as subjects of numerous books, documentaries, and of course, *The Conjuring*. While Ed Warren was not a priest certified by the Catholic Church in exorcism, he considered himself highly educated in all aspects of religious demons. He died at age seventy-nine in 2006. In his obituary, he was listed as one of only seven religious demonologists in the world. Through study, Ed Warren came to understand the hierarchy of Christian demons, and used this knowledge in his and Lorraine's supposed supernatural encounters.

One such event was with the Perron family in 1974. The Perrons, two parents and five daughters, had moved into a Rhode Island farmhouse in 1971. The house itself was rather unremarkable with white siding, an enclosed front porch, and an A-frame roof. Yet, in the three years before the Warrens conducted a séance, the Perron family reported all

Both Ed and Lorraine Warren claimed to have experienced supernatural phenomena as children. Ed lived in a haunted house where doors opened and shut, and Lorraine could see colorful auras around others starting at age nine.[2]

manner of ghostly disturbances, from a wandering female spirit dressed in eighteenth-century clothes, to shaking beds and other ear-splitting noises.

The Conjuring 2 is based on the 1977 "Enfield Haunting," in which Peggy Hodgson and her four children were said to be terrorized by a poltergeist for nearly eighteen months.[3]

Several members of the family, as well as Lorraine Warren, shared their experience with New Line Cinema to aid in the writing of *The Conjuring*. At the climax of the film, when demon-possessed Carolyn Perron (Lili Taylor) must be exorcised before she kills her children, Ed

Warren knows that he must perform the exorcism, as there is no time to waste. He is the only one who can speak and understand Latin! In his paper for *Social Research*, "Christian Demonology in Contemporary American Popular Culture," Armando Maggi describes how the film's climax mirrors the persecution of a witch:

> The exorcism of Carolyn Perron is staged as a private and dark representation. A single light bulb, dangling from the ceiling, lights up the area where the young mother is tied to a chair and covered with a sheet. On the one hand, we know that the sheet hides a victim who should not be held responsible for the attempted murder of her child. On the other, the words recited by the lay exorcist trigger such violent reactions in the possessed woman strapped to the chair and such dramatic disarray in the ritual space (for example, a flock of crazed pigeons circles the house and then shatters the window of the cellar in a scene reminiscent of Hitchcock's *The Birds* [1963]) that one cannot help but see these aggressive outbursts as originating from the woman herself, as if the exorcism were bringing to the fore her true, until then hidden, nature . . . The Warrens cover Carolyn with a sheet when, growling like a dog, she pins Ed Warren down on the floor and tries to strangle him, and then savagely bites a police officer's cheek. After restraining her, the lay exorcist begins to perform the ritual without being able to look her in the eye, which is unusual during an exorcism. Early modern treatises on witchcraft, however, recommended that the judge avoid making eye contact with the witch because of her evil eye. The exorcism of Carolyn in *The Conjuring* thus also evokes the trial of a witch.[4]

Like any movie, some excitement was added for cinematic effect, and in truth, Ed Warren didn't conduct any exorcisms during his career. But eldest daughter Andrea Perron (played by Shanley Caswell) told *USA Today* in 2013 that something just as unsettling happened at her family's farmhouse. The Warrens conducted "a séance (that) allegedly caused Carolyn Perron to be temporarily possessed, which Andrea claims she secretly watched. 'I thought I was going to pass out,' Andrea says. 'My mother began to speak a language not of this world in a voice not her own. Her chair levitated and she was thrown across the room.'"[5]

After the harrowing events in Rhode Island, Andrea Perron became an author. Her three-volume book series House of Darkness, House of Light (2011–2014) outlines her supernatural experience in what was called the "Arnold Estate" and how the séance led by Ed and Lorraine Warren ended her ghostly torment. While there are many skeptics of the Perron's story, Andrea Perron is quick to point out that there is an answer rooted in science for what occurred. "There is a scientific explanation [to what happened], we simply haven't discovered it yet."[6]

While the thought of a ghost trailing into your room and watching you sleep is sufficient fuel for nightmares, that wasn't enough for The Conjuring. A confused human spirit who can't find the way to their afterlife is a common trope in horror, and this frustration is often shown through moving objects and sorrowful wails in the attic. In The Conjuring and its sequels, The Conjuring 2 (2016) and The Conjuring 3: The Devil Made Me Do It (2021), gothic ghosts are usurped by more malicious entities. They work to inhabit living humans to do their bidding.

In The Conjuring 2, this is Valak, a demon using the ghost of an elderly man to gain entry into Janet Hodgson (Madison Wolfe). In The Conjuring 3, Arne Cheyenne Johnson (Ruairi O'Connor) is possessed by a woman simply known as "Isla the Occultist" (Eugenie Bondurant) who uses her powers to kill. Much like in the original film, Isla is a witch bent on destruction. She uses a Stregherian book of magic at her altar, which all must be destroyed in order to halt Isla from her evil machinations.

Stregheria is an Italian form of witchcraft, its name literally derived from the Italian word strega, meaning witch:

Stregha witchcraft goes all the way back before the Christian era because the Italian word for witch comes from the Latin "strix" period, meaning screech owl. In the Roman era, there was a belief that "striges" (which is plural for strix) were women who could transform with the help of magic into birds of prey. Pliny the Elder wrote about this belief, and the Roman author Apuleius described this transformation in his book The Golden Ass in the second century CE.[7]

Much like other facets of witchcraft, stregha has changed over the centuries, becoming influenced by Greek and Spanish cultures. The practices of stregha were revived in the 1970s by Italian Americans like author Raven Grimassi who, for his book *Hereditary Witchcraft: Secrets of the Old Religion* (1999), studied ancient Italian writings in order to learn the witchy ways of the original stregha. Thanks to those continuing the traditions, many of the tenets have stayed the same from ancient to modern stregheria, including the worship of nature gods and goddesses.

While Isla twisted her stregheria practice into possession and murder, the witch at the center of *The Conjuring* has a more personal torment. In that film, Lorraine has a vision of Bathsheba Sherman (Joseph Bishara) hanging from a tree in the Perron's backyard. Unnerved by this vision, she researches the history of the house and discovers that accused witch Bathsheba was said to have sacrificed her infant to the devil. She then hanged herself from the very tree that Lorraine had witnessed in her trance. While this backstory is equally fascinating and creepy, is it true?

The answer is . . . sort of. Like many aspects of *The Conjuring*, there is a seedling of truth that inevitably grows into a spooky, cinematic tree. Bathsheba Sherman was a real woman, who, indeed, lived on the land that the Perrons inhabited decades later. Born in 1812, Bathsheba was married to a farmer and was the mother of a son when accusations of witchcraft were thrown her way. The local legend was that a baby in her care died from a large sewing needle impaled in its head. This gruesome death led Bathsheba's neighbors to accuse her of sacrificing the infant to the devil. She was tried and "due to insufficient evidence a court found that she was innocent of any wrongdoing. Despite her name being cleared legally, the public was not convinced."[8] Unlike her film counterpart, Bathsheba did not take her own life. While she was forced to live in a community that villainized her, Bathsheba remained in the farmhouse and raised her son, Herbert. She died an old woman in 1885.

The notion of child sacrifice, while unsettling, has existed for centuries. In "Child Sacrifice, Ethical Responsibility, and the Existence of the People of Israel," Omri Boehm pointed to several examples from the Bible in which there is a story or at least a mention of the practice. Boehm also cited other sources in his pursuit of discovering how authentic child sacrifice truly was in ancient times:

Non-Biblical evidence for the existence of the myth of child sacrifice is given by Eusebius of Caesarea. Once again, this is only "secondhand" testimony and yet it seems that, with all due caution, we can trace the common pattern. According to Eusebius, Philo of Byblos attests that at a time of danger to the Phoenicians the leader would sacrifice his son to the gods in order to appease them. This leader would also circumcise himself and his close man for the same purpose. Eusebius likewise cites another testimony, giving an explicit explanation of the tradition, saying that at times of danger to a city or population, leaders would sacrifice their sons to appease the gods and save their people.[9]

During the "Satanic Panic" that swept the United States in the late 1980s, child sacrifice came back into the spotlight, but unlike Abraham in the Bible who was willing to kill his son for God, those accused of child sacrifice were believed, like Bathsheba, to seek the approval of Satan. A fervor, sparked by an episode of *20/20* (1978–) as well as numerous media articles and books had gripped the country. People believed that their kindly neighbors were conducting blood rituals of the occult. As outlined in *The New York Times*, lack of true evidence didn't necessarily stop innocent people from going to prison.

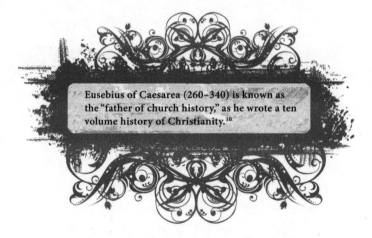

Eusebius of Caesarea (260–340) is known as the "father of church history," as he wrote a ten volume history of Christianity.[10]

Nearly two hundred people were charged with crimes over the course of the Satanic Panic, and dozens were convicted. Many defendants were eventually freed, sometimes after years. Three Arkansas teenagers who became known as the West Memphis Three were freed in 2011, almost twenty years after they were convicted of murders that prosecutors portrayed as a satanic sacrifice. In 2013, a Texas couple was released after twenty-one years in prison; they were later awarded $3.4 million from a state fund for wrongful convictions.[11]

More recently, witch doctors (a gender-neutral term in Uganda) who believe that children's deaths will bring about prosperity to an area plagued by poverty, drought, and disease, take matters into their own hands. In an interview with *USA Today*, Joel Mugoya, a traditional healer, explained this disheartening practice, saying "there is no food due to the ongoing drought, and some believe that this has been brought by ancestral spirits, so there is a high desire for people to conduct sacrifices so that they come out of this problem."[12]

The "West Memphis Three" were teenagers when they were convicted of killing three eight-year-old boys in 1993. Because of the teens' interest in heavy metal and their goth appearances, they were singled out, and a motive of child sacrifice was assumed. They were later found to be innocent of the gruesome murders.[13]

SECTION FIVE
COVENS

CHAPTER FOURTEEN
THE CRAFT

On its release, *The Craft* (1996) was given lackluster reviews. Yet, it cast a spell on a generation of girls. They rented this teen-witch horror film from their local video stores, finding catharsis in teenage characters who were so much like them: unpopular, misunderstood, and drawn to the darkness. *The Craft*'s popularity is on the rise as an interest in feminist horror and the occult has caused the next generation to become fans of the witchy four. A sequel, *The Craft: Legacy* (2020), was recently released to much attention.

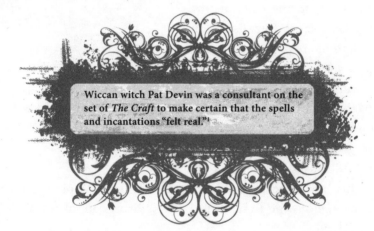

Wiccan witch Pat Devin was a consultant on the set of *The Craft* to make certain that the spells and incantations "felt real."[1]

In *The Craft*, Sarah (Robin Tunney) is a lonely, suicidal teen who has natural witch abilities. It isn't until she meets Nancy (Fairuza Balk), Bonnie (Neve Campbell), and Rochelle (Rachel True) that Sarah finds a coven where she can appreciate her given talents. Covens have long been a standard in witch stories. From Shakespeare to cartoons, witches are often depicted in groups, their proximity increasing their power. In *Vulture*, Angelica Jade Bastien describes the inherent distrust of covens, in *The Craft* specifically: "It taps into the primal American fear of female power, and what happens when women are alone together, forming hothouse bonds that seem to only occur in adolescence in ways that are both profound and contradictory, liberating and limiting."[2]

Soon, it is revealed to new girl Sarah (Robin Tunney) that this coven worships an entity known as Manon. As Nancy describes, "If God and the Devil were playing football, Manon would be the stadium that they played on; He would be the Sun that shone down on them." Manon is an ancient pagan-like entity that is in all things, present in nature, in energy. When the girls chant, it is for his power to flow through them. It was important to the filmmakers that they created a fictional pagan god that harkened back to known nature deities, while not being authentic. They wanted to both be respectful of the pagan religions, as well as not encourage teenagers to summon evil. Lofty goals, to be sure!

Paganism is a broad term, once used to describe any religion that was not Christianity. Modern paganism is rooted in many forms, often nature-driven like Wicca. Because many pagan religions are polytheistic, there are thousands of gods to choose from in the creation of Manon. He was most likely inspired by those deities who brought destruction, like Whiro, a Māori god of death who looks like a lizard and causes people to do evil

Morrighan is a Celtic pagan goddess warrior who in later Irish folklore became known as the banshee, a frightening entity who could foresee death.[3]

deeds. This link of serpents and death gods could explain the use of snakes throughout the film, and of course brings about images of the Christian devil.

Another familiar witch trope used in *The Craft* is that of offering blood in a ritual. There is animal sacrifice, as well as a scene in which all four girls must cut the heels of their hands and bleed into a chalice of wine, which they then drink from. In her dissertation on blood beliefs in early modern Europe, Francesca Matteoni discusses the vital link between blood and witchcraft:

Theories regarding the body played an important part in the complex of witchcraft beliefs which nourished the period of the trials and persecutions in Europe. The witch-body became the evidence of a dangerous power according to the mentalities of both common and learned people: the accusers, the judges, and the accused themselves. The body and especially its fluid, the blood, became the means of exchange between a supernatural force and the physical world.[4]

Tropes naturally inhabit the shadowy world of a horror film. From the beginning of cinema, films centered on witches have shared particular traits that moviegoers enjoy and expect—spells, familiars, good versus evil—though, over time, the narratives have shifted, as our collective understanding of witchcraft has blossomed.

The Harry Potter (1997–2007) series is one of the most banned sets of books in recent years, with Pope Emeritus Benedict XVI even sharing his thoughts on the popular young-adult books by J. K. Rowling in 2003, saying they "deeply distort Christianity in the soul before it can grow properly."[5]

In order to learn more about the intersection of modern film and witch, we spoke to Payton McCarty-Simas, author of the thesis "I Am that Very Witch: Gender, Genre, Psychedelics, and Abjection in the

2010's Witch Horror Cycle," winner of the Pat Anderson Prize in Film Reviewing, and nominee for the Andrew Sarris Award for Film Criticism from Columbia University.

Meg: **"First, can you tell us about your adolescent exposure to witches in the media? What were some of your favorite fictional witches?"**

Payton McCarty-Simas: "Being raised in Massachusetts, witches were in the culture—my parents got married at one of the witch churches in Salem and we were taught about the Witch Trials in school from an early age. On top of that, *Harry Potter* was at the height of its cultural relevance in the 2000s and, like lots of kids, I really wanted to be Hermione (Emma Watson). She was smart, ambitious, and she used her magic to do well in school on top of saving her friends. I rented *Kiki's Delivery Service* (1989) from Blockbuster constantly, too.

The most important witches for me personally though as a queer kid growing up were the Hex Girls from *Scooby Doo and the Witch's Ghost* (1999). Their magic, unlike the witches and wizards in *Harry Potter*, was deeply rooted in environmentalism and a tradition of American Wicca (they talk about *The Craft* in depth). Plus, they were super cool goth rock stars, and so queer-coded! I had such a crush on Thorn, the lead singer. That movie is interesting because it sets up a dichotomy where not all witches are inherently evil. 'Bad witches,' stereotypical old crones, use their black magic for power, but Thorn, the Hex Girls' hereditary witch, is a practicing Wiccan whose powers save the day for everyone. In contrast to other films I watched as a teenager like *The Craft*, in which Fairuza Balk's Nancy Downs, a girl who looked and acted a lot like one of the Hex Girls (and who I also loved!), gets violently institutionalized and punished at the end of the film, it was comforting to have a portrayal of a confident, young, powerful witch like Thorn who isn't pitted against her coven sisters and depicted as 'hysterical' or dangerous to herself and others."

Kelly: **"That's such a vital difference! In your thesis you point to the Carol F. Karlson quote, 'the story of witchcraft is primarily the story of women.' Can you give us an idea of what this means, particularly in the Western sense of the witch?"**

Payton McCarty-Simas: "Carol F. Karlson's book *The Devil in the Shape of a Woman: Witchcraft in Colonial New England* (1998) does an excellent job laying out the ways in which witchcraft as an idea or a label rather than a practice was wielded against women who didn't conform to heteropatriarchal standards of femininity. In Colonial New England for example, she points out that the vast majority of the people tried and murdered as witches were women who were viewed by their communities as threats to the dominant order, whether they wanted to own property, practice medicine, express their sexuality, or even take a more active role in Christian religious leadership. This fear of feminine strength and a desire for autonomy is so baked into the idea of being a witch that the *Malleus Maleficarum*, the classic Medieval witch-hunting text, often uses the word 'witch' and the word 'woman' interchangeably."

Meg: "Wow! I had no idea!"

Payton McCarty-Simas: "Scholar Nel Noddings theorizes that the representation of the witch as the ultimate anti-Christian is in part a strategy to make paganism, which often lauds fertility and eroticism as divine, as threatening rather than appealing to Christians. Eve, a figure of unbridled sexuality and appetite when paired with the serpent (a Pagan symbol), is depicted as catastrophically dangerous to humanity while the divinely guided Mary represents perfect purity even in childbirth because God caused her conception. Thus, sexuality is reviled only outside the control of male authority.

The work of nineteenth-century scholar Jules Michelet affirms this theory, directly holding Christian repressiveness and hypocrisy along with the cruel strictures of Feudalism responsible for the birth of the witch—women are for him not inherently evil but pushed into Satan worship by their circumstances. The witch as a figure comes to represent a concrete threat of feminine power specifically based in Pagan practice rather than the more amorphous danger of 'otherness,' be it sexual, racial, or otherwise, that characterizes most movie monsters. In terms of the films I discuss in my thesis, this trope becomes central. The witch characters' narratives in these films are fundamentally a tremendous struggle with—and ultimately rejection of—this Christian feminine paradox. Even though Christianity posits women as weak-willed and in need of male protection, inherent in the figure of the witch in this context, as in Eve before her, is the admission that women do hold tremendous power (sexual, reproductive, and otherwise) that must be controlled at the risk of destroying the fragile patriarchal order. More than anything else, that's the real threat witches represent: women set free from the constraints of heteropatriarchal morality."

Meg: **"You explore a rise in interest in subjects like the occult, spiritualism, and satanism throughout history. How has the trope of the witch changed over time? And how do world events play a role?"**
Payton McCarty-Simas: "This was one of my favorite subjects to research, and ultimately the conclusion I drew was that although the sociopolitical context changed over time, the threats witches and the occult represent to hegemonic power remain fundamentally the same. The characteristics attributed to witches in the West across history can be described simply as 'anti-feminine.' Accusations of witches spoiling crops, sickening babies and livestock, causing women to miscarry, and rendering men impotent were common in the Middle Ages through the colonial era—all details that paint a portrait of the witch as an anti-reproductive woman. At the same time, witches were depicted as potently sexual beings who seduce men, women, and animals alike, disrupting the Christian family structure through this other anti-feminine paradigm wherein sexual agency is masculinized, particularly for women past reproductive age.

These characteristics remain essentially the same as time progresses. Witches, mediums, and spiritualists in the Victorian era represented a threat to the scientific establishment with their 'irrational' beliefs and practices and sexually explicit acts of spirit channeling. Satanic Panic in the 1980s, too, represents a similar fear of 'perverse' sexuality and corruption of the youth. Since the mid-2010s, the rise of conspiratorial thinking and far-right political engagement has led to a similar moral panic in the form of the QAnon movement's claims that liberal politicians perform Satanist ritual abuses, pedophilia, and cannibalism of young children.

All three cultural periods I've just mentioned . . . are characterized by their moral conservatism, religiosity, and a resulting obsession with sexual transgression, from puritanism to Victorianism to Reaganism to Trumpism. Interestingly, though, contemporaneous with these historical unfounded accusations of witchcraft and satanism comes an equal and opposite rise in Wicca and satanist practice. Today in particular: according to a 2014 Pew poll, there are more practitioners of Wicca in the United States than ever before—comparable to the number of American Presbyterians! And that number is expected to keep increasing. This speaks to the way both Satan and the archetypal figure of the witch have been lauded by disempowered groups, particularly women, as the ultimate rebels who reject unjust authority, while representatives of traditional Christian values demonize them for the same reason. The fact that these two contrasting waves exist in tandem paradoxically heightens the apparent legitimacy of the perceived threat felt by conservative Christians and adherents to extremist or puritanical ideologies like QAnon."

Kelly: **"As you point out, the witch is the only really prominent female monster in horror cinema. Why do you think that is? And how do you feel about the portrayal of the witch as the 'bad guy'?"**

Payton McCarty-Simas: "Witches are unique in the pantheon of movie monsters because, unlike a zombie film or a vampire film, witch films can look like so many different things—the tropes aren't really fixed. Additionally, while people in the Middle Ages believed in vampires and werewolves as well as witches, only the witch has a fundamental basis in reality: those condemned as witches were real people. Before modern medicine became a regulated institution, these women's healing powers were viewed as magic and thus a real threat to the Christian notion of life and death as the sole purview of God. Because witches have always represented at their core the fear of female/feminine power . . . it makes sense that the most oft-represented female monster on screen would be a witch. This archetypal figure is so deeply ingrained in our collective unconscious, she makes for rich and variegated source material that eludes the hyper-formulaic structures of other monster films because hers is a uniquely ambiguous, abject threat.

Personally, as a queer person, my relationship to 'bad guy' characters is complex. From Hammer films (*Dracula* for example) to Disney (Ursula or Maleficent come to mind), villains are often queer-coded, degraded, and killed. I enjoy witches as 'bad guys' for the same reason they're presented that way in the first place: they tear down social hierarchies and scare the you-know-what out of people (mostly men and authority figures)."

Meg and Kelly: "Woohoo!"

Payton McCarty-Simas: "As Stephen King points out in his excellent book *Danse Macabre* (1981) . . . horror movies as a cultural form are fundamentally conservative in their worldview. They present the cultural fears and anxieties of their makers and let them run wild, ultimately eliminating them at their close. To quote King, '[t]he ritual outletting of these emotions [terror, disgust, anxiety, etc.] seems to bring things back to a more stable and constructive state again' once the credits roll. That being said, the joy for most horror buffs I think comes from those middle passages during which the villain is allowed their time to run amok. Who remembers what the campers' names are in *Friday the 13th* (1980)? There's no sequel without Jason, Freddy Krueger, Dracula, Pinhead, Jigsaw . . . the list goes on, but the heroes of these movies come and go with time. And, as we know from slasher franchises like *Halloween* . . . the resolutions offered in the final moments of most modern horror are far from definitive—behind each cozy conservative conclusion lies the potential for radical change. Plus, no other genre is as willing to reveal the fragility of our societal institutions in the first place, making horror movies a natural draw for people disenfranchised by those same power structures.

Of course, as with the queer-coded villains of old, the question of positive representation is always important, particularly with a villain fundamentally based in an archaic fear of powerful women. But, as I discuss in my thesis, this most recent cycle of 'witch films' troubles the traditional horror villain narrative and presents the witch as a sympathetic evil pushed to her limits by a heteropatriarchal system designed to deny her autonomy. The good guys in horror movies are almost always boring anyway."

Meg: "Ha! You're right, I would much rather play the witch than the final girl!"

Kelly: "**Can you explain your research on both the 1974 novel and the 1976 film versions of *Carrie*? I found that piece of your thesis fascinating.**"

Payton McCarty-Simas: "Carrie White is a really excellent vector for discussing most of the tensions at the heart of the cinematic witch figure. She's caught between two entirely contradictory sets of feminine conventions: her mother's conservative Christian belief that the female body is innately sinful, hypersexual, and abject, and the worldview presented by her teenage peers that the female body's sexuality is an asset to be flaunted and used for social capital. As the narrative progresses, these two versions of feminine selfhood clash inside the girl who struggles to overcome the internalized religious sexual repression instilled by her mother and embrace a 'normal,' yet equally strict and hierarchical, version of teenage girlhood.

Horror scholar Shelley Stamp Lindsay has described *Carrie* as 'not about liberation from sexual repression, but about the failure of repression to contain the monstrous feminine,' and this is true, but only to a point. Carrie White's internalized religious guilt is incapable of outweighing her desire for teenage sexual exploration and socialization—she embraces her powers in a notable rejection of the Christian moral doctrine at the heart of what makes the witch archetype threatening—and yet, she is still overwhelmed by the powers she possesses due to her feminine coded hyperemotionality, another archaic stereotype about women which comes from the same set of Christian ideals.

Carrie is exemplary of a wave of horror films in the 1970s that represented renewed cultural anxieties surrounding a weakening of patriarchal control over the feminine following the tremendous social repression of the 1950s. Stephen King himself acknowledges that his novel 'is largely about how women find their own channels of power, and what men fear about women and women's sexuality . . . an uneasy masculine shrinking from a future of female equality.' The thing that makes it most significant to my research however is the fact that, as Stamp Lindsay points out, *Carrie* makes real strides to overcome the conservative narrative conventions of the period.

Even after her death, Carrie's presence still haunts the film. The last scene shows one of her 'good girl' classmates (Amy Irving) having a nightmare wherein she goes to Carrie's (Sissy Spacek's) now burned house (she is a witch, after all) and sees Carrie's hand burst out of the ground to grab her. The monstrous feminine power Carrie represents, the power of the witch to overcome both the heteropatriarchal values of the church and, in a more modern context, the high school pecking order, is always threatening to break free inside the mind of even the nicest young women. Carrie then becomes an integral steppingstone for the 2010s cycle because the anxiety of that moment is exactly what these modern films agonize over and delight in."

Meg: **"Tell us about the modern witch horror film. How does that differ from the witch movies of our youth?"**
Payton McCarty-Simas: "Essentially, the witch horror films of the 2010s cycle (as defined by male auteur filmmaking from studios like A24) break dramatically with horror tradition because, instead of dying horribly in these movies, the witches get the last laugh. Films like *The Witch* (2016), *Suspiria* (2018), *Midsommar* (2019), and *Color Out of Space* (2019), come to a close with the totally radical and fascinatingly consistent image of a young woman smiling or laughing with complete ecstasy and abandon, reveling in her newfound supernatural powers as everyone around her dies horribly. Seeing this happen over and over again at the movies, I was dying to know why—and that's when I started my research. This new climax comes with a lot of new and interesting tropes and conventions such as a consistent use of psychedelics and psychedelic imagery that I argue is used to signify a narrative departure from the 'normal' heteropatriarchal world into the realm of an older, Pagan, feminine-coded space defined by its chaos, unfamiliarity, eroticism, and danger (and which parallels with the 1970s witch horror cycle). There are the references to flying ointment and ergot fungus in *The Witch*, the magic mushrooms and other potions in *Midsommar*, the color that alters reality in *Color Out of Space* (an underappreciated film, I think—more people should see it if they like the other movies I'm talking about!), and the list goes on.

Interestingly, this witch horror cycle that I was exploring seemed to be dying off just as I was writing about it in 2020 and another shift was taking place. *Gretel and Hansel* (2020) adopts the aesthetic sensibility of the cycle with its psychedelic imagery and Pagan symbology, but falls back on the conventional narrative tropes of an evil old crone and an innocent young girl who defeats her. Theoretically, the fact that Gretel (Sophia Lillis) herself possesses magic could be understood as a feminist intervention into the orthodoxy of the story, but to me it's a weak one. She poses no real threat to the dominant order and vows to keep herself in check (suggesting innate hysteria in need of controlling) and to serve only conventionally moral notions of 'good' with her powers. It could actually be seen as a symbolic reinstatement of the repression Carrie White and other young witch characters worked so hard to overcome in decades past.

At the same time, female-directed films like *The Craft: Legacy* (2020) and *Black Christmas* (2019) attempt to push back against some of the archaic tropes at play by depicting female witches as unambiguous forces of justice in the face of patriarchy. *The Craft: Legacy* in particular was a significant step for the subgenre because one of the main witches is trans (Lourdes, played by Zoey Luna), contesting the historical/mythic narrative that a witch's feminine power comes from her vagina. Still, these movies lack the real radicality and danger of the core films of the cycle."

Kelly: **"How do you see the future of witches in the media? Are they here to stay? Will they continue to revel in their powers, rather than be 'saved' from their own power?"**
Payton McCarty-Simas: "The media landscape has changed so much since I finished my thesis in the winter of 2020. I've continued to track witch movies as they've come out, and sadly, I think my suspicion that the cycle ended in 2020 was correct. With movies like *The Unholy* (2021) and *The Conjuring: The Devil Made Me Do It* (2021), witches are back to their old habits. In the former, the ghost of a witch burned during the Witch Trials (they posit that those innocent murdered women were in fact harbingers of evil) possesses a sweet young Christian girl only to be soundly defeated by a brave male protagonist. Similarly, in *The Devil Made Me Do It*, another witch from Massachusetts of the old crone variety sees her curse undone by the power of wholesome heterosexual love in a conclusion so shockingly retrograde people in the theater I saw it in were openly laughing. As with any film cycle, success and innovation almost always breeds opportunistic imitation that eventually waters down everything exciting about the originals until the cycle can't sustain itself anymore, but these new movies are interesting because more than that typical decline, they actually signal a backlash—a complete regression to

a previous mode of conservative representation. It's even more compelling when viewed in relation to the shift from a conservative to a democratic presidential administration, and I'm planning to continue tracking this new cycle in tandem with the political climate. Thankfully for us, though, since horror movies are almost always profitable, there's plenty out there to watch from the past half decade for people who really loved this cycle, and there are always people out there taking risks and making original, exciting, low-budget work in the horror genre."

Meg: **"Tell us about your current and upcoming projects?"**
Payton McCarty-Simas: "As I've been following the trajectory of witch horror in the 2020s, my interest has shifted into religious horror more broadly. So, I'm currently doing research for a piece on that as well as editing a piece I wrote previously about the evolution of the psychedelic horror film. I'm also editing lots of short films and I'm in the process of

The Brothers Grimm fairy tale about two children finding a witch in the woods has been inspiration for numerous horror films, like the South Korean adaptation *Hansel and Gretel* (2007), 2013's action-horror *Hansel and Gretel: Witch Hunters*, modern retelling *Hansel and Gretel* (2013), *Hansel vs. Gretel* (2015) in which Gretel has become a witch, and many more![6]

writing my second feature screenplay! It's a supernatural horror story about a folk choir instructor coming to terms with her childhood trauma in a small town in Maine and I'm very excited about it!"

Kelly: "So many intriguing projects! We can't wait to read, or watch, what comes next!"

We owe a huge thank you to Payton McCarty-Simas for such insightful answers. The concept of "cycles" within the witch subgenre was something new to us that we were eager to explore. From *The Craft* to *The Craft: Legacy*, it's evident that popular culture, politics, and the slow shift of gender roles play integral parts in horror films.

CHAPTER FIFTEEN

ꓢUSPIRIA

In 1845, British essayist Thomas De Quincey published the first sections of his *Suspiria de Profundis* in *Blackwood's Magazine*. What was planned to be thirty-two parts, though not finished before De Quincey's death in 1859, *Suspiria* was a fantastical meditation on memory and experience through the prism of psychedelic drugs. De Quincey's most famous essay, *Confessions of an English Opium-Eater* (1821), focused on the nature of drugs and addiction. Many consider him the first purveyor of addiction literature in the Western Hemisphere.

While *Suspiria de Profundis* never reached De Quincey's ambitious number of planned essays, the ones published and uncovered posthumously are considered literary gems. One popular example is *Leavana and Our Ladies of Sorrow*, an imaginative telling of De Quincey's dreams of Roman goddess Leavana during his time at Oxford.

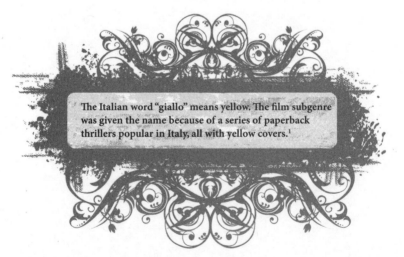

The Italian word "giallo" means yellow. The film subgenre was given the name because of a series of paperback thrillers popular in Italy, all with yellow covers.[1]

Within *Leavana and Our Ladies of Sorrow* is the harrowing description of her companions, a trio of witchy sisters: Mater Lachrymarum, Mater Suspiriorum, and Mater Tenebrarum.

Three sisters they are, of one mysterious household; and their paths are wide apart; but of their dominion there is no end. Them I saw often conversing with Levana, and sometimes about myself. Do they talk, then? O, no! mighty phantoms like these disdain the infirmities of language. They may utter voices through the organs of man when they dwell in human hearts, but amongst themselves there is no voice nor sound; eternal silence reigns in their kingdoms. They spoke not, as they talked with Levana; they whispered not; they sang not; though oftentimes methought they might have sung, for I upon earth had heard their mysteries oftentimes deciphered by harp and timbrel, by dulcimer and organ. Like God, whose servants they are, they utter their pleasure, not by sounds that perish, or by words that go astray, but by signs in heaven, by changes on earth, by pulses in secret rivers, heraldries painted on darkness, and hieroglyphics written on the tablets of the brain. They wheeled in mazes; I spelled the steps. They telegraphed from afar; I read the signals. They conspired together; and on the mirrors of darkness my eye traced the plots. Theirs were the symbols; mine are the words.[2]

These supernatural sisters became the inspiration for Italian director and screenwriter Dario Argento in the creation of his Three Mothers trilogy of horror films which consisted of *Suspiria* (1977), *Inferno* (1980), and *The Mother of Tears* (2007). Like in De Quincey's essay, these formidable witches ruled over separate cities, their unspoken prominence bringing death and destruction to Freiburg, Germany, New York City, and Rome. Before *Suspiria*, Argento was known as one of the first filmmakers of the Italian giallo movement, a subgenre rising in popularity in the 1970s and 1980s of murder mysteries rife with gory action. While *Suspiria* is not considered giallo because of its paranormal subject matter, Argento returned to the bloody thrillers with popular films like serial killer fare *Tenebrae* (1982).

In *Suspiria*, American dance student Suzy Bannion (Jessica Harper) comes to Germany hoping to study ballet, only to find out that the dance academy is plagued by brutal deaths. One such fatal encounter happens to the studio's blind piano player, who made the tragic mistake of angering

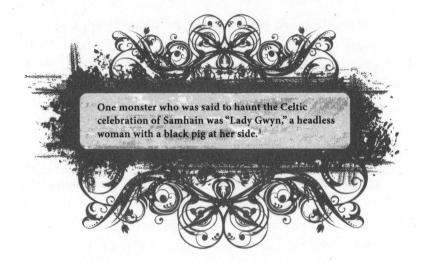

One monster who was said to haunt the Celtic celebration of Samhain was "Lady Gwyn," a headless woman with a black pig at her side.[3]

the witches. Alone in a plaza at night, with only his guide dog at his side, Daniel (Flavio Bucci) senses an evil presence. Suddenly, his beloved dog turns on him, savagely clawing and biting at Daniel's throat until he dies. The horror is palpable as film critic Calum Cooper explains:

> Not only is it so unique a scene for a horror film—as it inverts the often exploited feeling of claustrophobia into one of agora-phobia—but it encapsulates the sinister nature of witchcraft best. It is not in a witch's appearance that makes them so scary, but in the seemingly endless potential of their magic. By being able to tinker with even the tiniest details with a flick of the fingers or the incantation of a word, virtually anything can be bent to the spellcaster's will. In presenting many of its most horrific scenes in unconventional spaces, settings, or even lighting, *Suspiria* conveys just how chilling being at the mercy of witchcraft can be, especially when their motives seem purely for the sake of being evil more than anything else.[4]

Just like in De Quincey's essay, the witches of *Suspiria* need only to whisper their intent in the wind or perhaps snap their fingers to turn a loving dog against his owner. This complete hold on all aspects of nature is scarier than any wart-faced archetype of witch.

Thirty-nine members of Heaven's Gate were discovered dead on March 26th, 1997. Each deceased person had exactly five dollars and seventy-five cents in their pants pocket.[5]

I (Meg) became intrigued by this particular pull on the natural world and how witches, good or evil, might interact with their environment. Soon, I discovered the concept of elemental manipulation. In its most rudimentary form, these are the four most commonly used elements: earth, wind, air, and fire. Surely *Firestarter*'s (1984) Charlie McGee (Drew Barrymore) would've been considered a witch for fire manipulation if she'd lived before the Industrial Revolution! Fire, too, is prominent in the climax of *Suspiria*. Once the most formidable witch, Helena Markos (Lena Svasta), is killed by Suzy, the entire coven bursts into flames. Since their powers over nature have been subverted, they are now victims of their own fiery rage. An example of a fictional witch with sway over water is Countess Von Marburg (Lucy Lawless) of the CW series *Salem* (2014–2017). Von Marburg was called the "water witch," as she used rain to slow her enemies and nearby water to drown them. This harkens to De Quincey's aforementioned description of the three witchy sisters, "by changes on earth, by pulses in secret rivers."

Do current practicing witches seek to manipulate the elements around them? In Wicca, sometimes known as pagan witchcraft, a modern religion that is based in occultism, the answer is yes. Wiccan author Lisa Chamberlain explains the vital nature of elements to her religion:

For thousands of years the Elements have been considered the basic building blocks of the universe, and they are inherent to the basic principles of witchcraft. Wiccans and other witches honor, recognize, and participate with these core energies in both religious ritual and magic. By attuning to and working with the magical qualities of the elements, both individually and combined, witches are able to manifest positive changes for themselves and others, fostering a deeper spiritual connection to the natural world.[6]

Wicca focuses heavily on nature. Wiccans celebrate cycles of the moon and sun and respect nature in all its forms. While the witches in *Suspiria* are involved in nefarious plots to murder, Wiccans are quick to note that they practice white magic to only make the world a better, more peaceful place.

One popular Wiccan tradition is Samhain, a pagan festival that was born in the beautiful land of the ancient Celts. Wiccans have revived this celebration in recent decades. Believed to be when the barrier between the worlds of the living and dead is at its most thin, the Celts chose the end of October to usher in the darkest half of the year. About five weeks after the Autumnal Equinox, those in ancient times would recognize the longer nights as a time to reflect on their dead loved ones as well as celebrate the fall harvest:

Autumnal equinox, two moments in the year when the Sun is exactly above the Equator and day and night are of equal length; also, either of the two points in the sky where the ecliptic (the Sun's annual pathway) and the celestial equator intersect. In the Northern Hemisphere the autumnal equinox falls about September 22 or 23, as the Sun crosses the celestial equator going south. In the Southern Hemisphere the equinox occurs on March 20 or 21, when the Sun moves north across the celestial equator. According to the astronomical definition of the seasons, the autumnal equinox also marks the beginning of autumn, which lasts until the winter solstice (December 21 or 22 in the Northern Hemisphere, June 20 or 21 in the Southern Hemisphere.)[7]

This date of October 31st remained Samhain for centuries, and for most, has turned into the jack-o'-lantern and trick-or-treating good time known as our favorite holiday, Halloween! Wiccans, on the other hand, have learned from the Celts in the traditional ways to honor the festival. They incorporate Wiccan rituals into the ceremonies, often including fire, an element used by the Celts in bonfires, and even an ancient game in which men throw flaming logs at each other!

In *Suspiria*, Suzy is instantly thrown into the intrigue as a vicious double murder happens on the night of her arrival.

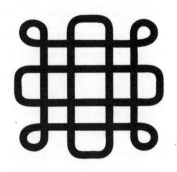

The looped square symbol for Samhain is the Bowen knot—also called St. John's Arms and Gorgon's loop.[8]

Strange occurrences keep coming, creating a sense of dread for both the students in the academy and the viewers watching this hyper-colored film. Soaked in deep reds, vibrant greens, and blues, *Suspiria* is a sumptuous horror treat, its visuals are often pointed to by critics more than its characters or plot. In fact, it is not clear, until Suzy does some research of her own, that witches are even the antagonists of the film. Once she speaks with Frank Mandel (Udo Kier) and Professor Milius (Rudolf Schundler), Suzy comes to understand who the immortal Helena Markos is, and how the coven works. She must kill the witch in power, and then her coven will perish.

The concept of a leader dying and their followers thus facing a dismal fate exists in the real world. One example is in the beaching of whale pods. While a single beached whale is not out of the ordinary, when an entire pod is found dead on land, scientists wonder. Many studies have been conducted and there are numerous factors at play. There is one hypothesis, the "Sick Leader Theory" that suggests whales or similar creatures like dolphins may beach themselves because they have followed a sick leader. Their imperative to follow is more important than their biological need to survive.

A group of long-finned pilot whales stranded themselves on a beach in Scotland. One of them was an old, diseased male, who may have been the "sick leader." This male may have purposely veered into shallower water because he wasn't fit enough to stay in the depths, says Andrew Brownlow of Scotland's Rural College in Inverness, who led the investigation into the strandings. "One of the theories is that animals will strand themselves when they are very weak because they don't want to drown," says Brownlow. He suggests there might be "something very deep in the terrestrial mammalian core that fires up when they are in extremis." Alternatively, it might be that the whale knew it was ill and fled to the shore to protect its relatives from the sickness—only for them to follow.[9]

This same tendency toward suicide has been seen in countless human examples, namely when a cult leader dies, leaving their followers without guidance. Many follow their leaders in death immediately, like in the case of Heaven's Gate and the Jonestown Massacre. For Benjamin Nathaniel Smith, a member of the neo-Nazi cult, The Church of the Creator, he followed in the footsteps of the leader he idolized, Ben Klassen. Klassen killed himself in 1993 after the death of his wife. This was no surprise as he had written a book entitled *The White Man's Bible* (1981) which called suicide "an honorable and dignified way to die for any . . . of a number of reasons, such as having come to the decision that life is no longer worthwhile ."[10] Six years later, Klassen's disciple, Smith, took his own life, but not before he targeted random minorities on the street and gunned them down. Two of his victims, Ricky Byrdsong and Won-Joon Yoon, were killed. This sort of suicidal extremism has found its place in many political movements, creating an opportunity for leaders to manipulate their followers like a witch might cast a spell on the elements.

The coven of witches in *Suspiria* certainly didn't choose to perish at the hands of Suzy Bannion. They were working all their witchy powers for the sake of evil, in order to continue their dark reign over Germany. Though, when they abandoned all humanity and became Helena Markos's vicious army, the commitment to their coven was their eventual, fiery downfall.

CHAPTER SIXTEEN
THE WITCHES OF EASTWICK

It was in a college course on American short stories that I (Meg) had the first occasion to read John Updike. The assigned 1961 story, "The A&P," centered on a teen boy's view of three women visiting a grocery store, left me wondering what, if any, was the subtext? And if there was a deeper meaning, did I care? After that single story I decided to avoid Updike.

A decade or so later, I was confronted with a rather confusing dilemma. For our podcast, *Horror Rewind*, Kelly and I chose to rewatch and analyze the 1987 horror comedy *The Witches of Eastwick*. This film had been a staple of my childhood, its scenes, both scary and hilarious, forever etched across my cinephile heart. So, when the opening credits began, and the words "based upon the novel by John Updike," crawled across the screen I felt a sense of dread in the bottom of my belly. No! In my avoidance of Updike, I hadn't had the chance to learn that *The Witches of Eastwick* was *his* novel, written in 1984. I was now walking a precarious tightrope. Could I still love the film, while rejecting Updike and his work? Turns out, yes, with no problem at all, I can.

At the time of the book's release, Nina Baym wrote a scathing article for the *Iowa Review* in which she echoes my sentiment on the author: "If Updike despises his witches' victims, he doesn't much like the witches either, though their total immersion in self and sex he finds refreshing, really womanly. Skimming rapidly from one consciousness to another, he prevents the depiction of individual minds among these witches, though he expends pages on the nuances of their bodies."[1]

To be fair, many reviewers enjoyed the novel, and the author himself described it as feminist. Some scholars agreed while others balked at the idea. Wherever you find yourself in that debate, what we will focus on in *this* book is the film.

Directed by George Miller, most recently at the helm of the hugely successful *Mad Max: Fury Road* (2015), *The Witches of Eastwick* has a

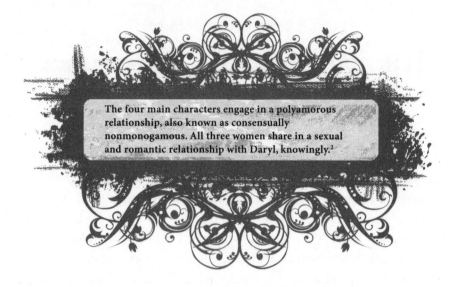

The four main characters engage in a polyamorous relationship, also known as consensually nonmonogamous. All three women share in a sexual and romantic relationship with Daryl, knowingly.[2]

dream cast: Susan Sarandon, Michelle Pfeiffer, Cher, and Jack Nicholson. The three women are divorcees, living in the small, fictional New England town of Eastwick. While they have had jobs, marriages, and kids without any sense of magic, something has changed. While the timing is never explained, this burgeoning ability to cast spells is strengthened when they are together, in their coven of three.

After a little dalliance with weather control, the women unknowingly conjure a stranger into Eastwick. They take turns describing their dream man, and he soon arrives in the storm. This bringing forth of a man (or a demon, as we come to find out), is also known as evocation. Not only associated with witches, evocation of spirits or demons is part of religions like Daoism and Shintoism. Spiritism, a movement founded in the nineteenth century by Allen Kardec, also explores the conjuring of spirits. Kardec believed that a spirit existed after one's death, and worked to prove this through science and research. Unlike spiritualism, spiritism is focused on reincarnation. Modern believers use Kardec's books on spiritism to hone their philosophy:

Spiritism consists of two parts: one of these, the experimental, deals with the subject of the manifestations in general; the other, the philosophic, deals with the class of manifestations denoting intelligence. Whoever has only observed the former is in the position of one whose knowledge of physics, limited to experiments of an amusing nature, does not extend to the fundamental principles of that science. Spiritist philosophy consists of teachings imparted by spirits, and the knowledge thus conveyed is of a character far too serious to be mastered without serious and persevering attention.[3]

Witches in the media conjure all kinds of bad entities, like the coven in *The Witches of Eastwick*. At first, Daryl (Jack Nicholson) seems to be a positive change in the humdrum lives of the witches, but as their powers and confidence grow, the women recognize his malevolence. A major clue is when he uses his demonic powers to scare each woman with her worst fear.

For Jane (Susan Sarandon) it is the fear of aging. For a horrific moment she goes decades into the future, her skin wrinkled, her bones twisted. While it is normal to feel anxiety about the changes old age wreaks on our bodies and minds, people with gerascophobia have an intense panic at the thought. According to the International Center on Aging, this phobia has many factors, including the fears of disability, loneliness, dementia, and death. They have recommendations to battle these persistent thoughts:

> If you have disabilities, even if you have memory loss, you will continue to value your ability to do what you want to do and to interact with other people in a meaningful way. Some of your interests and passions for life will remain the same. New interests and activities can appear that you cannot even imagine, things that you will love and that will make you happy that your life plan allows you to do them. Be kind to your future self! Plan ahead your future years and work with others to create a society where older adults are valued and respected.[4]

Next, comes Daryl's wrath directed toward Alexandra (Cher). She has expressed her distinct fear of snakes, also known as ophidiophobia. She awakes in her bed, absolutely covered in the slithering creatures. It's a writhing mass of different species, causing her to scream and become scared for her life. While we don't know how intense Alexandra's fear truly is, some people have such debilitating ophidiophobia that at the mere flash of a picture of a snake they can react with dizziness and nausea. Actor Matt Damon is one such sufferer, according to his *We Bought a Zoo* (2011) costar Scarlett Johansson. She told *People* magazine that the scariest moment working in the film was "watching Matt Damon cry like a baby and rock back and forth when the snakes were spread all over the set."[5]

In a 2001 Gallup poll, snakes were the number one fear of Americans, beating out public speaking and heights.[6]

While Jane and Alexandra had a few moments of fright, Sukie (Michelle Pfeiffer) was targeted with the most ire from Daryl the demon. Her biggest fear is physical pain, or algophobia. Daryl uses a piece of fruit as a sort of proxy to Sukie, smashing and biting it, as Sukie feels every infliction of pain. She is even hospitalized for the lingering devastation to her body. While most of us don't have to endure supernatural attacks, the human experience inevitably involves pain. In an interesting study outlined in the medical journal *Pain*, there is proof to suggest that one's anxiety over feeling pain actually makes the hurt more substantial.

To study these relationships, researchers began by administering various psychological tests to their human subjects to determine individual anxiety levels. The first, the Anxiety Sensitivity Index questionnaire, measured the subjects' levels of anxiety about different physical sensations like heart palpitations or stomach aches. For example, it measured whether they might interpret a bad stomachache to mean they had stomach cancer or had just eaten a bad pizza for lunch. The second, the Fear of Pain Questionnaire, measured fear levels for different types of physical pain ranging from a paper cut to a broken neck. Participants were then subjected to varying levels of pain applied to their forearms with a heat probe—never hot enough to damage tissue but enough to be uncomfortable. While the pain was administered, researchers collected images of their subjects' brains. Finally, researchers correlated the questionnaire scores with brain activity in specific regions of the brain. The study found a high correlation between the Fear of Pain Questionnaire and the right lateral orbital frontal cortex, an area of the brain that when activated may reflect attempts by fearful individuals to evaluate and/or regulate responses to pain.[7]

After the witches have recovered, they realize they need to work together to rid Eastwick of Daryl. They trick him into thinking they all still want to be in a relationship with him, and then use a spell book titled *Malefico* to conduct a series of spells to vanquish him. In Italian, *malefico* means "harmful" which seems apropos for the thrilling climax of the film.

> In hoodoo, followers believe in a positive sort of spiritual possession, rather than the typical scary demon inhabiting our bodies. It is considered an honor to be possessed by Ioa, a spirit who guides your dreams and connects you to the spirit world.[8]

As soon as Daryl is out of sight, the three witches use wax to sculpt a poppet. Poppets have a long history in witchcraft and are often associated with hoodoo dolls. The principle is the same: the poppet doll is used much like Daryl's piece of fruit, a proxy to represent the person that the magic is being performed on. But a hoodoo doll is closely associated with hoodoo spiritual beliefs, and therefore is not a witch's poppet.

One of the oldest spells known that involves a poppet was recorded on a stone tablet in Mesopotamia. It instructed the victims of dog bites to rub clay on the wound, use the clay to sculpt a dog, and then, after saying a few magical words, leave it outside facing the sun. Since there was no modern medicine, this spell may have been the only option for a sufferer!

In Graeco-Roman practice Kolossoi were small dolls used for spells—often curses. They were named for a victim, melted or broken, and then taken to a cemetery where the spirits of the restless dead could be asked to give the person casting the spell what they desired. This technique probably traveled to Britain.[9]

A vital aspect in the poppet spell is the use of personal items, even body fluids or parts, in the construction of the doll. Fingernail clippings, saliva, or in the case of the film, hairs from Daryl's brush, are added to the wax poppet. If one is sewing a cloth poppet, they would stitch these things inside. While the witches have reason to treat the Daryl poppet with violence, not all poppets are used in this way. "Healing poppets are common and convenient. You can make one to stand in place for a

The practice of tar and feathering was first documented in 1189 in England, when Richard I outlined the punishment for thieves. They were to have boiling pitch, or resin, poured over their heads and feathers attached for further humiliation.[11]

distant loved one who may need healing. You can perform healing spells or Reiki on the doll, stuff it with healing herbs, anoint it with a healing oil and it will send the energy to its human equivalent."[10]

That's all well and good, but Alexandra, Jane, and Sukie need to scare or even kill Daryl in order to be free from his control. They impale the wax with long needles, causing him pain. They even "tar and feather" him, make him vomit uncontrollably, and cause his body to fly and jerk. Furious, Daryl turns into his true form, a massive beast, focused on killing the women. They get the better of him, throwing his poppet into fire. While Daryl may be a demon, he is no match for this powerful coven.

A rare look at witches forming their powers a bit later in life, *The Witches of Eastwick* is a story of female friendship and power that no man, not even a handsome devil, can tear apart.

CHAPTER SEVENTEEN
DEATH BECOMES HER

It is not an exaggeration when I say that *Death Becomes Her* (1992) was one of the most profoundly influential films of my (Meg's) childhood. We didn't go to the movie theater often as a family, so my memory of watching this horror/science fiction/comedy mashup is an indelible touchstone of my youth. My parents, older brother, and I shared popcorn as we watched, shocked by the quality of the visual effects. Thirty years later, I still remember the cardboard cutout of Helen Sharp (Goldie Hawn) awaiting us at the cinema. She wore a shocked expression, her red hair gleaming. As an eight-year-old drawn to the dark side, I was most intrigued by the hole in the cardboard Goldie's abdomen. I could stick my hand through it! Before I even entered the darkened theater, I knew I was going to fall in love with *Death Becomes Her*.

In the original mock-up for Helen's shotgun blast to the stomach, SFX artists made it much more gruesome, with spine and guts showing! But filmmakers wanted a less realistic injury to match the comedic tone of the film.[1]

Female driven, snarky, and unexpected, the film directed by Robert Zemekis gave me something I was craving as a fledgling film fanatic: women who were complicated and even downright nasty. By then I had grown tired of the woman as the innocent victim trope in horror and beyond, and had found myself drawn to biting, calculating women like *Death Becomes Her*'s Helen Sharp and Madeline Ashton (Meryl Streep). They joined a small assemblage that I was forming of beloved female baddies like Ruth Patchett (Rosanne Barr) in 1989's *She-Devil*, as well as women with questionable morals like those in *Clue* (1985).

At the center of *Death Becomes Her* is a familiar plight of women, one that resonates in many of the films, books, and series covered in this book: our appearance, and more specifically our obsession with youth. Witches are often reduced to their looks, like the simplistic dichotomy of *The Wizard of Oz*'s Glinda the Good Witch (young and beautiful) and her sister the Wicked Witch of the West (ugly and haggard).

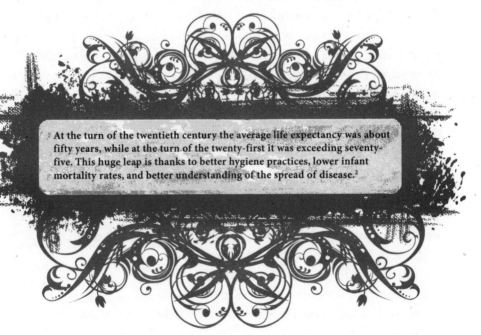

At the turn of the twentieth century the average life expectancy was about fifty years, while at the turn of the twenty-first it was exceeding seventy-five. This huge leap is thanks to better hygiene practices, lower infant mortality rates, and better understanding of the spread of disease.[2]

In our book *The Science of Women in Horror*, Kelly and I touch on "hagsploitation," a cinematic movement in which actresses like Bette Davis fully leaned into their advanced age, their wrinkles used as an ode

to the monstrous. Yet on the whole, women, especially those in entertainment, like Madeline and Helen in the film, are expected to maintain
an almost supernatural element of youth. Using whatever scientific
advancements are available, from surgery to supplements, Hollywood
and much of the world is fixated on turning back the clock. This is on
display early in the film when Madeline goes to a clinic, begging for
the latest treatments to make her appear youthful. In the background
we see a horrifically comic scene of a woman going through what looks
like torture. She is strapped to a rotating gurney, IVs of blood going in
and out with all matter of poisons. She is a modern-day Frankenstein,
willing, like Madeline and Helen, to be an amalgam of plastic parts to
keep up with an impossible ideal.

Bruce Willis plays against type as Dr. Ernest Menville, a plastic
surgeon with no proverbial spine who almost allows the strong women
in his life to kill him. While at the beginning of the movie it seems like
Ernest is the reason Madeline and Helen are fighting (he leaves Helen to
marry Madeline), it becomes clear as the story unfolds that he is merely
a pawn in a decades-long game that the women constructed. He, like
all the men in their path, is used as a social status symbol, and a way to
hurt the other. Because of our obsession with youth and appearance, the
film argues, we have pitted woman against woman.

Women's relationship with our own
appearance is not a new phenomenon,
and neither is plastic surgery. And while
there is an obvious toxicity when obsession with appearance goes too far, the
need to appear "normal" and feel com-
fortable in one's own skin is a universal
necessity. Writings from Sushruta, an
Indian man of medicine who lived in the
sixth century, describe both skin grafts
and nose reconstruction. A skin graft is
a surgical procedure used commonly in
modern times when a piece of skin from

According to the Aesthetic
Society's national databank, the
most common plastic surgery in
2020 was liposuction, followed
closely by breast augmentation,
and thirdly, abdominoplasty,
also known as a "tummy tuck."[3]

one part of the body, or sometimes from a cadaver, is used to mold or
construct somewhere else on the patient's body.

Sushruta's treatise provides the first written record of a forehead flap rhinoplasty, a technique still used today, in which a full-thickness piece of skin from the forehead is used to reconstruct a nose. At that time, patients in need of that procedure generally included those who had lost their noses as punishment for theft or adultery. Today, surgeons use skin grafts to restore areas that have lost protective layers of tissue due to trauma, infection, burns, as well as to restore areas where surgical intervention has created a loss of skin, as can happen with melanoma removal. Some grafts include blood vessels and muscle, such as in reconstructive breast surgery.[4]

Christianity and spiritualism trumped science through the Middle Ages, although one vital procedure, to repair a cleft lip, was developed in the tenth century. A cleft lip is a rather common birth defect in which the baby in utero is forming its face between the fourth and seventh week of gestation, and the tissue at the lip does not fully touch from both sides. This can leave a small to large opening on the upper lip. It can also affect the nose and palate.

What eventually led to plastic surgery's sudden ubiquity was the atrocities of World War I:

Military physicians were required to treat many extensive facial and head injuries caused by modern weaponry, the likes of which had scarcely been seen before. These grave injuries necessitated brave new innovations in reconstructive surgical procedures. Some of Europe's most skilled surgeons dedicated their practices to restore their countries' soldiers to wholeness during and after the war. It was in fact around this time that surgeons began to fully realize the potential influence that one's personal appearance could exert upon the degree of success experienced in his or her life. Because of this understanding, aesthetic surgery began to take its place as a somewhat more respected aspect of plastic surgery.[5]

By the early 1990s when characters Helen and Madeline were desperate for treatments, plastic surgery had become a billion-dollar industry, promising to maintain youthful skin on the face and a tight, toned body.

Yet, science has its limitations. Aging is the great equalizer, our bodies do it from the moment we are born until we are laid to rest. What the women of *Death Becomes Her* get, of course, is more than they bargained for. Lisle Von Rhuman (Isabella Rossellini) provides both Helen and Madeline with a potion that not only reverses aging but gives them eternal life. Like many witches before her, Von Rhuman offers what sounds like a great deal. But the reality is much darker. With eternal life comes eternal challenges, especially after Helen gets a shotgun blast to the belly and Madeline breaks her neck falling down the stairs!

Von Rhuman is stunning, confident, and mysterious, living in her castle-like mansion in Beverly Hills. She embodies everything the "hag" witches we have come to know do not possess, though her motivations are quite familiar: snag unsuspecting victims, take what they can give you (in this case, money), and let them be tortured by the consequences.

In our pursuit to understand the impact of *Death Becomes Her* today, we came to find that it has caused quite a lasting ripple effect in the drag community. To learn more we had the pleasure of talking to two popular drag queens, actor and comedian Jinkx Monsoon, winner of season five of *RuPaul's Drag Race* (2009–), as well as filmmaker and event producer Peaches Christ, who has appeared in such films as *Milk* (2008) and developed the stage show "Drag Becomes Her" (2019).

Kelly: **"This book is about witches in the media. Can you tell us some witchy inspirations from your childhood? Who was the first witch to catch your attention?"**

Jinkx Monsoon: "I have been obsessed with witches my entire life. I can remember so many witches I worshiped as a young person. There was Magicka D'Spell in the animated show *Ducktales* (1987–1990) as well as Morgana from the show *Darkwing Duck* (1991–1992). I remember being terrified by Anjelica Huston in *The Witches* (1990) and also wanting to be her. A very special character from my childhood was Queen Mab, played by Miranda Richardson, in the made-for-TV movie *Merlin* (1998). And of course, the Sanderson Sisters from *Hocus Pocus* (1993), played by the holy trifecta of Midler, Najimy, and Parker. There were countless witch inspirations I found throughout my life—but these are some of the earliest and most nostalgic examples."

Meg: "I loved all these! We definitely grew up in the same era!"

Peaches Christ: "Like so many, it was absolutely the Wicked Witch of the West in the movie *The Wizard of Oz* (1939). She and her winged monkeys were perhaps my first horror obsession. I loved everything about her! To me, Margaret Hamilton's performance is the pinnacle of witch performances. She's perfection as the witch and she's fabulously hilarious and mean as Miss Gulch. I've never stopped being obsessed with *The Wizard of Oz* and it often shows up in my stuff."

Kelly: **"She is the most iconic witch *ever*! There has been a feminist reclaiming of the witch archetype. Why do you think that is? Do you find witches empowering?"**

Jinkx Monsoon: "The feminist reclaiming of the word 'witch' is one of the main reasons I talk about witches and witchcraft as publicly and as often as I can. I think we're living in an age where we are tearing down many long-standing falsehoods that have plagued our social consciousness for too long. One of those things is the demonization of the witch archetype. The misogynist depiction of women as witches and witches as villains is a tired and transparent trope of the patriarchy to keep powerful women suppressed. I feel like this reclamation of the word 'witch' and the open practice of witchcraft is very empowering, and just one more way we can take the power away from our oppressors."

Peaches Christ: "I love that feminists have reclaimed the witch archetype. Feminists have been given such a bad rap and for what? Wanting equality? They've been called witches and bitches and Nazis and all sorts of shit so I love that there's this sense of 'well, if believing women are equal to men makes me a witch, I'll be a witch.'"

Meg: **"I had a visceral reaction to *Death Becomes Her* as a young girl. I've watched it many times and it still resonates. What has been your personal experience with the film?"**

Jinkx Monsoon: "I first saw *Death Becomes Her* when I was five or six years old, and I was head over heels, crazy in love with that film. I would have my mom rent it for me every time she went to the video store. Of course, Lisle Von Rhuman was a fantastic witch—but Madeline Ashton and Helen Sharp are also witches in their own right—once they've essentially sold their soul for eternal youth and beauty. Besides the magical elements, this movie is wrought with fierce femininity, as our main characters battle it out within the confines of an image-obsessed world that shames women for the 'egregious' act of aging."

Peaches Christ: "I too loved it from the moment I saw it. I remember being a senior in high school when it came out and I just couldn't understand why it wasn't a smash hit. I felt like my friends and I had discovered this incredible gem of a movie, but somehow it just wasn't catching on. It had an all-star cast, a star director, big budget special effects, and yet it wasn't really a hit. I'm so glad that over the years it has found a real devoted audience of obsessed witches over the years who really 'get it.'"

Kelly: **"In a *Variety* article about how *Death Becomes Her* has found new life in the LGTBQIA+ community, Jinkx, you spoke about how one aspect of that is due to this notion of Helen and Madeline being outcasts seeking power. Can you elaborate? And Peaches, do you concur?"**

Jinkx Monsoon: "I think anything that displays disenfranchised people gaining power and finding a way to be fabulous in spite of the odds being against them will resonate with the queer community. In Madeline, we see a barely talented actor who made it big enough to become accustomed to a lifestyle that she is cast out of once she's considered past her prime; it very much exemplifies the Hollywood system of propping someone up, then tossing them aside once they're all used up. In Helen, we see a woman who was made invisible because of her image and the way she presented herself. Both women use magic to take control of their situations, reclaiming their youth, beauty, femininity, and sexuality—with the help of a magic potion. I honestly can't think of a plot that would appeal to drag queens (specifically) more."

Peaches Christ: "I definitely concur. I think that's a common theme in films that queer people connect to. Queer audiences relate to being used and tossed aside by people in power and for sure both Helen and Madeline are in this camp—they're women who dared to get fat or grow old and so the potion is, in a sense, a revenge fantasy. We also love that they join forces and unite in the end—that's very satisfying for the outsiders in the audience."

Meg: **"Tell us about 'Drag Becomes Her.' What is your role in the creation of this show and how has the pandemic affected the process? When will we get to see it?"**

Jinkx Monsoon: "'Drag Becomes Her' is the creation of Peaches Christ, inspired by the universal love that most drag queens have for the original film. The script was written by Peaches, the original music was parodied and written by BenDeLaCreme, and I enthusiastically star as Jinkx Monsoon playing Meryl Streep, playing Madeline Ashton. (It's very meta that way.) In the drag-centric, stage production of the movie, we zero in on the moments of the film that speak most to the LGBTQAI+ community and present them very much through a drag queen's lens. The pandemic prevented us from doing our third remounting of the show, but the film is timeless, and I like to think so is the live show based on the film. As soon as we can reschedule it, we will."

Peaches Christ: "I first created the show with myself and my friend Heklina in the lead roles many years ago and then updated it to star Jinkx and DeLa in 2018 when it premiered in San Francisco. I'm the writer, director, producer, and I play Isafella. I love this show and always have so much fun when I do it. In 2018 we took it to Seattle, Portland, Los Angeles, and then to the UK in 2019 for a sold-out run in Manchester and London. We were supposed to bring it back in 2020 but we all know how that went. Jinkx and DeLa are truly family so anytime we get to all work together is wonderful."

Kelly: "We will be watching closely for the chance to see it!"

Meg: **"Peaches, could you tell us about your film *All About Evil* (2010)? As horror fanatics we are so intrigued! It seems like a love letter to the genre."**

Peaches Christ: "It's my first and only feature film to date and I'm really proud of it. The movie is a dark comedy gore film that stars Natasha Lyonne as a murderous filmmaker. The movie was really inspired by my love of old movie theatres and grindhouse filmmakers like Herschell Gordon Lewis and Doris Wishman. I'm excited that it's being rereleased for its tenth anniversary and will be streaming soon. Also, Severin is doing a big BluRay with extras."

Meg: "Yay! I was hoping you'd say it would be available soon!"

Kelly: **"Jinkx, are you a horror fan?"**

Jinkx Monsoon: "I am absolutely a horror fan. Mainly because I love magical realism, which is why I'm less into slashers and more into supernatural themes. I think there is also an inherent camp factor in most horror films. There's almost a drag nature to the overly stylized death scenes that you expect from the best horror films. While I love the classics for the camp factor, modern horror, when done well, has become a profound expression of social commentary."

Meg: **"If you had witchy powers (and maybe you already do . . .) what is the *first* thing you'd do?"**

Jinkx Monsoon: "I must clarify that I am indeed a practicing witch, and I feel like I am actually quite content with the powers I already harness, and the balance of my life. But if I'm being candid and honest . . . I would love to be able to levitate. I can't help but think of all the ways I would enhance my live shows with the ability to levitate! And can you imagine if I got a wind machine in the mix? I just keep imagining myself as Agatha Harkness (Kathryn Hahn) in her final battle with Scarlet Witch (Elizabeth Olson) in *Wandavision* (2021–). Wasn't that scene amazing?"

Meg and Kelly: "Ah-mazing!"

Peaches Christ: "I guess I'd use them to help address things like world hunger, climate change, war, and economic disparity. I'd try to make the world a better place. I'm sorry it's such a depressing answer!"

Kelly: "No! That is perfect, what a great witch you would be!"

Meg: **"What projects are you working on? And how can we follow you?"**

Jinkx Monsoon: "You can follow me on my socials (@thejinkx on Instagram, @jinkxmonsoon on everything else. Be sure to subscribe to my podcasts if that's your thing (available anywhere you listen to podcasts) and start with the Pam Grossman interview where we discuss the tenets of modern witchcraft. You can also check out my YouTube channel for all the latest self-produced digital content I've been creating at home."

Peaches Christ: "I'm working on my *Midnight Mass* (2021) podcast that I cohost with Michael Varrati. He and I are also collaborating on some movie projects. You can keep with me on all of the social media channels as well as my site peacheschrist.com or my immersive theatre site terrorvault.com."

Just like after a viewing of *Death Becomes Her*, we couldn't stop smiling from the fabulous answers from Jinkx Monsoon and Peaches Christ! Their insight into their love of the film further enhances our own devotion to a few of our favorite witches in cinema.

Death Becomes Her won a visual effects Academy Award for its impressive depiction of its "Hollywood zombies" and was the first film to use silicone as an animatronic skin.[6]

SECTION SIX
WITCHES ON TELEVISION

CHAPTER EIGHTEEN
A DISCOVERY OF WITCHES

A Discovery of Witches (2011) has the premise of a professor spending hours doing research in a dusty library, discovering her witch powers, plus a love story. Count me in! I (Kelly) read the books and dove into the television series when it premiered. Deborah Harkness, author of the All Souls trilogy (2011–2014), was inspired to create this story in part due to her own past as an historian. Where research ended, she saw the possibility for fictional stories to fill in the gaps.

> What inspired me was to try and think about how I could write a fairy tale for grown-ups that was about these fantastic creatures living among us. I explain the world from inside their communities. I think in that way it's an approach that would make sense to other historians of science because what we do is study the systematic ways the people in the past looked at the world and their place in it.[1]

Real life inspirations for the mysterious series and the character of Matthew (Matthew Goode) were based in history. For example, Matthew Roydon was a sixteenth-century English poet and friend of playwright Christopher Marlowe. "He was a spy for the queen, yet we don't have firm information on where or when he died. He was mentioned in Marlowe's accidental death inquest, but nobody knows where he was buried. It was as if he vanished."[2]

In the first episode of *A Discovery of Witches* (2018), Dr. Diana Bishop (Teresa Palmer) is lecturing on alchemy. Alchemy is a medieval chemical science and speculative philosophy whose aims were the transmutation of base metals into gold, the discovery of a universal cure for diseases, and the discovery of a means of indefinitely prolonging life. Its most famous endeavor was the creation of the philosopher's

Alchemists used jargon in their recipes and historians have determined that unlikely sounding ingredients were shorthand for more regular things. Dragon's blood refers to mercury sulfide and the black dragon represents a form of powdered lead.[3]

stone, which purportedly could turn base metals into gold and give immortality. Women were instrumental in the history of alchemy, including a woman called Mary the Jewess. She is credited with inventing heating and distillation processes that are still used in chemistry today. Although it was seen as related to magic in the past, more modern views of alchemy are forgiving of these now outdated beliefs.

The mid-eighteenth century was the beginning of the end for alchemy. By 1818 and the publication of *Frankenstein*, which portrays the mad scientist as basing his monster project on the work of famous alchemists, the discipline was considered to be solely the domain of crackpots, quacks, and con artists.[4]

Like all science, as we learn and discover new things, former ideas can be disproven. Neuroscientist Manuel Brenner said:

> In alchemy hides the germ of a protoscience that would later become the modern scientific method . . . I think there is an important lesson to be learned from alchemy. It shows us that many good and promising ideas (namely that you could run experiments to figure the world out, and that the world was to be understood through observation), combined with the wrong methods, can lead many intelligent, creative people, and even the most brilliant minds of their periods genuinely interested in acquiring knowledge, to keep running experiments and finding results that appear absolutely nonsensical from our point of view.[5]

Diana references the Ashmole book as a palimpsest. What is the history of such books? A palimpsest is a page of a book or a scroll that has been reused, perhaps several times, to create new writing.[6] Parchment was expensive, so previous writing was scraped or washed off with milk and oat bran. Over time, the previous writing would start to peek through. There are several famous examples of surviving palimpsests, including the complete text of the Institutes of Gaius (the first student's textbook on Roman law) and the only known hyper-palimpsest, the Novgorod Codex, where countless texts have left their traces.

Diana is a time walker in the books and television show. We've previously written about time travel in our book *The Science of Stephen King* (2020) but this got us thinking about witch walks. What are they and what is involved? Intuitive witch walks are a meditative practice that can help you get in touch with nature. It's recommended that you follow your intuition about direction and feel the energy of your surroundings. Look for signs in nature, it could be wildlife or plants, and see how they are speaking to you. Becoming aware of your surroundings, whether in nature or a city, can help you find answers to questions you haven't thought to ask.[7]

The witches in *A Discovery of Witches* make and write spells. What is the proper way to do this? First, it's important to realize that words have power behind them. What we say can affect ourselves and others. Patti Wigington recommends that you discover very specifically what your intention is. Next, decide what physical components you need for the spell. It could be traditional things like candles or herbs, but it could also be specific to your goal. Third, decide on timing. To some witches, casting spells during a specific phase of the moon is important. For others, it doesn't matter as much. Fourth, write the words to your spell and last, put it into motion.[8] Wigington also recommends that doing the work, whether it be sending out resumes for a job search, or meeting new people if trying to find a life partner, are important.

In many movies and TV shows people are shown burning sage to cleanse a space. What are the benefits of this? Burning sage, or smudging, is a practice of some Indigenous groups. It is seen as purifying, as it is a known antimicrobial. Smudging can be used to clear out spiritual impurities, insects, and pathogens. Sage is also used in some rituals to get in tune with the spirit world and it may have a basis in science. "Certain types of sage, including salvia sages and white prairie sage,

Channeling or speaking to spirit guides is seen in the scientific community as a phenomenon called "automatism—automatic behavior over which an individual denies any personal control … Today we recognize that automatism is a form of dissociation, an altered state of consciousness, in which an individual is capable of speaking or acting without awareness of deliberately doing so."[11]

contain thujone. Research shows that thujone is mildly psychoactive. It's actually found in many plants used in cultural spiritual rituals to enhance intuition."[9] One 2014 study showed white prairie sage as an effective remedy for anxiety and depression.[10] If you plan to burn sage make sure to do the following: open a window to allow smoke to escape, set your intention, light the end of your sage and blow it out quickly, allowing the smoke to linger on the parts of your body or the space you want it to, then collect the ash in a ceramic container. Always make sure to totally extinguish the smudge stick and store it in a dry place.

There are so many parts of witchcraft to explore, and we had the opportunity to talk to practicing witch, Ben Robinson, about his life and experiences.

Kelly: **"What was your first experience hearing about witches or witchcraft?"**
Ben Robinson: "I didn't have the memory of witches or witchcraft when I was little but my first 'magical' experience was at a place called Enchanted in Barnes, Wisconsin [a campsite]. I've learned in my adult life that it was my first step into magic. It was actually my Chosen (I shouldn't say my chosen deity—she chose me). I would wander away to be in the woods to collect rocks, gaze into the pond, and bond with the moon. When you are a witch, you don't have to follow a deity. You call upon the forces of nature itself. You work with them; you don't necessarily worship them."

Meg: **"When did you realize that you were a witch or wanted to become a witch? And how did you go about learning this?"**
Ben Robinson: "I always knew I was different. I didn't really have a name for it. I've always had an attraction to nature, and I've had a communion with cats, for instance. My first book to read about this was *To Ride a Silver Broomstick* (1993) by Silver RavenWolf. I read her series and that's what got me into Wicca to begin with. I believe everybody needs to have some kind of a sense of being a part of something greater than oneself. Whether it's a spiritual path or being a part of the community. Meeting the love of my life is what I needed to turn myself around."
Kelly: "We're so happy for you!"

Ben Robinson: "It was worth waiting for! I know that that sounds cheesy! He's a witch, too. He was in a very abusive relationship, and they tried to pray the gay away. He lost his spirituality so we're rediscovering it together."

Meg: **"What area of witchcraft do you feel most proficient in?"**

Ben Robinson: "I have this strong sense of knowing things. I've always had this since I was a kid. It's claircognizance, it's like a muscle you've got to work."

Ben told us about using a pendulum to receive answers from his spirit guide, learning to read runes, and we finished the interview with a single card tarot reading. The card that came up was "Strength," indicating "inner strength, courage, bravery, confidence, compassion, taming, control, overcoming self-doubt."[12] Ben also mentioned that tarot readings can be meditative and relaxing, helping you to focus on things you may not be aware of.

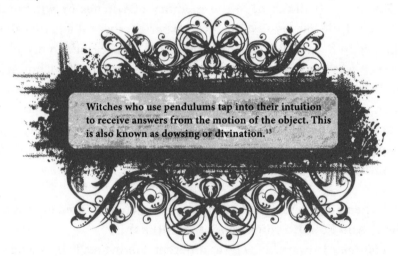

Witches who use pendulums tap into their intuition to receive answers from the motion of the object. This is also known as dowsing or divination.[13]

The Discovery of Witches travels throughout history and teaches the viewer about real people, books, and concepts. As Medievalist Sarah Durn said, "To medieval alchemists, the world was indeed a place of magic, of creatures, and phenomena they couldn't understand. Perhaps our contemporary world of science and technology could do with a touch of magic, of the unknown, of the alchemist's quest for enlightenment."[14]

CHAPTER NINETEEN
BUFFY THE VAMPIRE SLAYER

Although the Scooby Gang's main pursuit on *Buffy the Vampire Slayer* (1997–2003) was to vanquish said creatures, there was always plenty of room for more mythical beings to inhabit Sunnydale. It seemed a natural fit that one of Buffy's (Sara Michelle Gellar's) best friends, Willow (Alyson Hannigan), would come into those powers. Her journey to witchcraft began slowly, then grew out of control over the course of the series.

Not everyone was a fan of how Willow's portrayal of becoming a witch was handled. Jessica Mason noted in her piece for *The Mary Sue*: "they took a faith and practice that was about women finding their power and connection, and they made it into a metaphor for how power can corrupt a woman if she wants it for the wrong reasons. And also drugs? It was bad. Willow's magic was almost always associated with her most negative character traits and her poorest choices."[1] Those powers and poor choices came to fruition in an episode of the sixth season of *Buffy* in 2001 when Willow steals her girlfriend Tara's (Amber Benson) memories.

A study from 1990 to 2008 concluded that Wicca grew tremendously over this period. From an estimated 8,000 Wiccans in 1990, they found there were about 340,000 practitioners in 2008.[2]

"Once More, With Feeling" is the musical episode of the series that shifted several characters' plot lines going forward. Horror and musicals may not seem like a natural combination but I (Kelly), for one, am a huge fan. There's a killer musical number in *Scare Me* (2020), an adorable love story crossed with zombies in *Anna and the Apocalypse* (2018), and several horror stage musicals that haven't been made into movies (yet), including *Evil Dead the Musical* (2003), *Carrie the Musical* (1988), and *The Toxic Avenger* (2008). *The Shining* opera premiered in Minnesota in 2016 and had a sold-out run. One critic said,

> This musical version of Stephen's King's novel about the remote Colorado hotel that turns a man into a homicidal maniac succeeds at nearly everything it attempts. It marries the highest of art forms with one of the most suspect in popular culture. King's story is opened up and realized in new ways, and opera actually looks fun for a change. That's because composer Paul Moravec comprehends the situation at hand. His music is not overly serious or, as it might have been, camp, though it has elements of both. Moravec gets that most people know this story from the iconic, 1980 film starring Jack Nicholson, so he speaks the language of movie music—the shrill violin, the pulsing tempos of rising emotions, the bent note meant to warn audiences that things aren't what they seem.[3]

Unfortunately, neither Meg nor I were able to snag tickets to the opera but hopefully it'll be restaged someday. (Fingers crossed!) Besides overly enthusiastic musical theatre fans like myself, is it possible for people to burst into song and dance numbers unwillingly? There are several medical conditions that can cause involuntary vocal outbursts including

> Legends state that dhampirs, creatures that are the result of a union between a vampire and a mortal human, were normal members of the community. But dhampirs of paternal vampire descent could see invisible vampires and practice sorcery. They often started careers as vampire hunters which would be practiced for generations.[4]

Tourette's Syndrome, dementia, and schizophrenia. Alcohol and drug use can also be contributing factors to singing seemingly unwillingly. Klazomania is a condition first reported in 1925 and consists of excessive shouting. This condition is directly connected to encephalitis lethargica, a disease that attacks the brain, causing the victim to be motionless and speechless for a time. Symptoms include a high fever, lethargy, and a sore throat. Between 1915 and 1926 encephalitis lethargica ran rampant and more than one million people contracted it, nearly half of them died. Those who survived reportedly never returned to their previous state of health.[5]

> Being blackout drunk is a type of memory loss. "In medical terms this memory loss is a form of temporary anterograde amnesia, a condition where the ability to form new memories is, for a limited time, impaired. That means you can't remember a stretch of time because your brain was unable to record and store memories in the first place."[6]

How about breaking out into dance? Surprisingly, over ten centuries there were documented reports of dancing mania, also called choreomania or dancing plague, that occurred mainly in Europe. Groups of people would dance erratically until they collapsed from exhaustion or injuries and the reason behind it is still not understood. It was speculated that populations suffered from a mass psychogenic illness which is "symptoms affecting members of a cohesive group, originating from a nervous system disturbance involving excitation, loss, or alteration of function, whereby physical complaints that are exhibited unconsciously have no corresponding organic aetiology."[7] One common theme with the dancing mania incidents; they all seemed to happen during times of hardship. This was seen as a symptom of shared stress and trauma. How were people treated? Some were exorcised (if it was believed they were possessed by a demon) while others were isolated. One theory for the cause of dancing mania was ergot poisoning. Ergots grew during flooding and affected rye and crops. Ergotism caused hallucinations and convulsions in people. Whatever the cause, the next time I see people break out into a dance number in public I'll think about it differently!

Willow stealing Tara's memories on *Buffy the Vampire Slayer* was a pivotal moment in their relationship. It felt like a violation of trust on the deepest level and an unfair advantage on Willow's part to try to avoid conflict. How does memory loss actually affect people in real life? In 1998, Kelly's dad Bob was in a car accident with two fellow teachers. On their way to school, they hit a patch of black ice and the car they were riding in rolled. Nicole Baso Cuchna, a special education teacher in the carpool, was thrown into the highway on the other side of the median. We talked to her about her experience.

Kelly: **"I remember getting the phone call finding out that you and my dad had been in a car accident. What do you remember about that day?"**

Nicole Baso Cuchna: "I have no recollection of that day. The first person on the scene was a semi driver. He spotted my yellow jacket and thought that I was a piece of a snowmobile. When he got out of the semi, he realized I was a person. He then repositioned his truck so I would not get run over. I was fortunate that a doctor and nurse were some of the first arrivals. The doctor knew that they could not help me in Virginia, [the closest town] so he called the hospital and instructed them to get Lifelink ready to pick me up and bring me to St. Mary's Hospital in Duluth.

According to what I have been told, they had to use a defibrillator a couple of times in the ambulance on the way to the hospital. When I arrived at the hospital, Bob and I were put into separate bays. Bob told my mom that he heard me adding and subtracting numbers out loud like a calculator. As this was the end of the grading quarter at school, the night before the accident, I was calculating student averages for the report card. My brain must have been in calculator mode. I was bundled up and airlifted to St. Mary's in Duluth."

Meg: "It's so interesting how our brains work!"

Nicole Baso Cuchna: "My dad was the first person to arrive at the hospital to see me. I was in the emergency room when he arrived. The doctor pointed to my dad and asked me if I knew who this was? I responded, 'Dad.' An hour and a half later my mom and fiancé arrived. I was still in the emergency room. The doctor again asked me if I knew who these people were. I responded, 'no.' So within an hour and a half, I lost all knowledge of who my dad was and did not know my mom and fiancé. Later that evening I was moved into intensive care and still did not know who anyone was."

Meg: **"Tell us how your memory loss and amnesia manifested. What things or people did you remember? What things did you forget?"**

Nicole Baso Cuchna: "My brother and my two cousins (who are like sisters to me) came to Duluth the next day to see me. I did not know who they were. They would ask me questions and I thought I was answering them appropriately, but in reality, I had aphasia and left out many words of the

"About 40 percent of people aged sixty-five or older have age associated memory impairment—in the United States, about sixteen million people. Only about 1 percent of them will progress to dementia each year."[8]

conversation. After two days in the hospital, I started to call my mom "Mom." I wonder if I really knew her as a mom, or did I call her mom because that is what the doctors and nurses referred to her as? After about a week, the doctor came into my room and pointed to his watch, and asked, 'Do you know what this is called?' I responded, 'It has something to do with time.' Then he took out his keys and waved them in front of me and asked, 'Do you know what these are called?' I responded, 'It has something to do with your car.' He looked at my mom and said, 'We need to talk about rehab.'"

Kelly: **"That must have felt so strange! How did your memory come back over time, or not?"**
Nicole Baso Chuchna: "A neurologist had a meeting with my parents and fiancé. He told them, 'The person you knew as Nicole died on the highway. The new Nicole will be very different. She will never teach again and will never be able to learn new things due to the traumatic injury to her brain.' My mom responded, 'I don't think I like you very much.' He responded, 'I don't care. This is not a popularity contest. You will have to accept the reality of your daughter's brain injuries.' We started the tour of Miller Dwan [the rehab facility] and saw where I would attend physical therapy, occupational therapy, speech therapy, etc. It was during the walk through the rec area that I spotted a piano. I immediately walked to the piano, sat down, and started playing a song from memory. I still did not know anyone from my family or anyone else who came to visit me, but I remembered how to play the piano. I had taken piano lessons from kindergarten through college. When college friends and coworkers came to visit, I knew who they were and my associations with them. When my high school friends came to visit, I did not know them. To this day, I do not remember events from my childhood or high school. I look at videos, pictures, and yearbooks from that time in my life, and I recognize that it is me, but I have no memories of that time."

Meg: **"I know you postponed your wedding after the accident. Tell us about getting to know your fiancé again."**
Nicole Baso Cuchna: "After months of therapy and having an IQ test to meet state requirements to return to teaching, I moved back to Eveleth

toward the end of April and resumed my teaching duties. My fiancé, who was working in Fargo at the time, would drive to Eveleth to visit me. I did not remember him or our relationship. I knew he was a good person, but that is all I knew of him at this time. It was like having a stranger in my apartment. My mom came to visit me, and I told her that I did not want to marry Joe because I did not know him. It was a hard time for all of us. My mom suggested we all get some psychological help. Our sessions were individual therapy, therapy with Joe and me, and therapy with my mom and me. Through this therapy, I learned that it's okay not to remember every detail of my childhood. I can still have a happy life. Joe learned that if he wanted me in his life (even if only as a friend), he had to quit pressuring me and accept me as who I am. My mom learned to put the picture albums and videos away and stop asking me what I remember.

After the therapy, Joe would occasionally call me as a friend would and did not come every weekend to visit. As school ended, I was offered a job closer to home. I felt that this was a good option for me as it would help me with my memory. During that summer, I kept busy doing things with friends and family. Joe was included in those activities. I love baseball. Joe played baseball for the town team. I would go to every game and hang out with players and friends after the games. He would call and ask me on a date to dinner and sometimes a movie. Our relationship grew stronger. He had taken back the first ring and bought a different ring. By the time he asked me to marry him the second time, I had fallen in love with him again. We had our dream wedding in September surrounded by friends and family."

Kelly: **"That's incredible! How has your memory loss affected you long-term?"**

Nicole Baso Cuchna: "When people suffer a traumatic brain injury, you cannot come through it and be exactly the same as you were before. People ask my mom if I am different. Different does not always mean bad, peculiar, or odd. But I am different in that I like different foods than I did before the accident. I have a different taste in clothing. Because of the accident, I have become a stronger and more determined person. I went back to school to pursue my master's degree in school counseling. I feel that because of the experience I went through due to the accident,

I am a better counselor. I am able to recognize the frustrations and pain that some of these children are feeling because of their disabilities. There are so many things that are not known about brain injuries and memory loss. A brain injury is invisible to the eye. As a mother of three active girls ranging in age from sixteen to twenty-one, working full-time, sometimes it is easy to get overwhelmed.   To cope with my brain injury, I have the largest daily planner and the most sticky notes that you can imagine. Technology is a lifesaver: emails to help remember, notes on my iPhone, timers on my iPhone, etc. My life is very happy. I am blessed with three beautiful girls, a loving husband, a successful career, great friends, and a caring family."

Thank you to Nicole Baso Cuchna for sharing such a powerful and personal story with us about her memory loss! May we all appreciate every day and the memories we will make.

What are some ways to prevent general memory loss not caused by injury? The Mayo Clinic recommends being physically active, which increases blood flow to your brain, as well as staying mentally active. Playing games or doing something brain stimulating help your brain stay in shape. They also recommend being social, getting enough sleep, and staying organized. Good tips for us all!

CHAPTER TWENTY
CHILLING ADVENTURES OF SABRINA

Sabrina the Teenage Witch first appeared in comic books in 1962. Created by George Gladir and Dan DeCarlo, *Sabrina* followed the titular character using her powers in secret and dealing with the day-to-day issues of a teenage girl. Sabrina existed in various iterations including more comics, animated cartoons, and a television series premiering in 1996 starring Melissa Joan Hart which ran for seven seasons.

A favorite character on the show was the talking cat, Salem, who was a familiar. Voiced by Nick Bakay, Salem delivered many one-liners as a witch cursed to spend one hundred years as a cat. What are witches' familiars? They are household pets who see and know all and are considered to be guides, confidants, and protectors. In European folklore, familiars could appear as humanoid figures or shapeshift into various forms. They are most often cats but can be any animal that appears to the witch or is gifted to them. According to folklorist Icy Sedgwick:

In 1604, James I introduced a new Witchcraft Act that included "the occult rituals of diabolic witchcraft." This made working with evil spirits a capital offence. The act also referenced

Could a cat talk? Several species or groups of animals have developed forms of communication which superficially resemble verbal language. These usually are not considered a language because they lack one or more of the defining characteristics: grammar, syntax, recursion, and displacement.[1]

familiars, believed to be "the witches' helpful demonic companions." The Act really made an effort to clarify types of witchcraft.
It also turned communicating with spirits and practicing magic
with body parts into capital offences. The latter certainly veers
close to the divinatory practice of necromancy. By the time we
reach the witch hunts of the seventeenth century, these familiars
were more popularly referred to as "imps." They manifested as
small animals and a witch's power could pass to another person
through the familiar.[2]

In the 1997 book series *His Dark Materials*, each
character has a daemon; physical manifestations of the
soul put in place to help their human counterparts.

The latest version of Sabrina, *Chilling Adventures of Sabrina*
(2018–2020), like its predecessors, explores more topics than just being
a witch. It tackled relationships, family, fate, and self-love.

The Spellman family runs a mortuary out of their home and that
got us wondering more about the science behind this profession. We
interviewed Norcorafatimah Armstead (Cora), Kelly's former student at
Lake Superior College, who is now a funeral home director at Kozlak-
Radulovich Funeral Chapels in the Twin Cities in Minnesota.

Meg: **"Could you give our readers a little bit of background about
yourself?"**
Cora Armstead: "I was born on the small island of Penang on the west
coast of Malaysia. I came to Minnesota as an infant and received dual

citizenship. Our industry involves constantly learning, especially about different cultures and their funeral rites. It's so intriguing and not a single day is the same as the last. There are 'slow' seasons (in our case, summertime) and 'busy' seasons. Death is incredibly unpredictable but whether a family comes to our funeral home wanting an immediate cremation with no services or a full traditional funeral service with a casket and a burial, all my colleagues treat every single family with care and compassion. Being a funeral director is such a delicate yet honorable profession."

Kelly: **"Tell us about pursuing mortuary science and the training that it entails."**

Cora Armstead: "Most people who enter the mortuary science field have some familiarity with the industry. The majority of my classmates were born into families who owned a funeral home and were becoming funeral directors to continue the business that their families had started. People like me who have never come from a funeral home background found ourselves in the field due to an experience with death. Whatever the reason is, the field is definitely not for everyone. It is a lifelong sacrifice; long hours, eleven to twelve hour days, and constantly being on call.

The path to entering mortuary science differs from state to state. In Minnesota, most funeral directors complete two years of general courses, either obtaining their associate of arts degree or completing general courses at the University of Minnesota. Following that, most students transfer to the University of Minnesota's medical school to pursue a bachelor of science degree in mortuary science. During the one to two years of studies, the University of Minnesota sends students on two clinical rotations to separate funeral homes. Each rotation lasts approximately a month. During this time, the student works under a licensed funeral director that is designated as their 'preceptor' or 'mentor.' The University provides a list of items that must be completed for credit. This list includes a set amount of hours that must be spent working at the funeral home, a certain number of funeral arrangements, embalmings, funeral services, transfers of the deceased from the place of a death to the funeral homes, and graveside services. These clinical rotations are done prior to graduation and are unpaid. This allows students to gain experience and insight into the daily routine of a funeral home.

The next step is to take and pass the state and national boards. The national boards consist of two exams: the sciences (which pertain to the science and procedures of embalming) and the arts (which consist of the psychology and duties of a funeral director). In some states, being a 'funeral director' and 'embalmer' are two separate licenses. A funeral director makes funeral arrangements with families for their lost loved ones and assists with funeral and graveside services, while an embalmer takes the deceased into the funeral home's care and embalms the deceased (embalming is the chemical treatment of a body in order to temporarily preserve the breakdown that occurs after death).

Following passing the boards, the graduate is given a funeral directing intern license. The mortuary science graduate must complete a year-long internship (or 2,080 hours) at any funeral home of their choosing. Then, one is able to apply for their funeral directing license. After receiving your license, the learning doesn't stop there. This industry is constantly changing and growing, and the learning is lifelong."

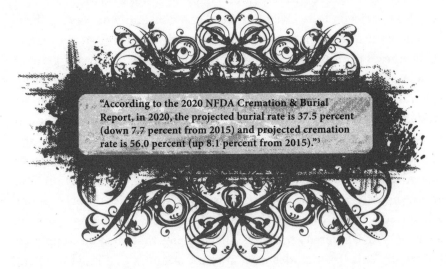

"According to the 2020 NFDA Cremation & Burial Report, in 2020, the projected burial rate is 37.5 percent (down 7.7 percent from 2015) and projected cremation rate is 56.0 percent (up 8.1 percent from 2015)."[3]

Meg: "Did you have a favorite course or one that you found the most interesting during your studies?"

Cora Armstead: "My favorite course that I took during my two years at the mortuary science program was definitely embalming laboratory. In this class, a small group of students are in a lab with decedents that have been donated through the Bequest program for medical study. My passion for the funeral industry began in the restoration of the deceased and I gained much satisfaction in getting hands-on experience with embalming, restoration, cosmetizing, and dressing the deceased.

A major aspect of funeral service that really supports families on their journey through grief is accepting that someone has truly passed. Viewing a loved one who is deceased plays a vital role in allowing someone to accept the reality of a passing. It is so important for families to say goodbye to their loved ones, and the deceased should appear as natural and peaceful as possible. As funeral directors, our job is to recreate that natural state."

Kelly: **"What do you think is a misunderstood part of being a funeral director?"**

Cora Armstead: "There seems to be a stigma that funeral directors (other synonyms are 'undertakers' and 'morticians') spend their time in a dark basement tending to the deceased. In reality, a large fraction of our day is made up of social interaction. Meeting with families at the place of death to take the deceased into our care, having families come to the funeral home to make funeral arrangements (which can last between one to two hours), assisting funeral services and graveside services at cemeteries. There is a lot of emphasis in school with understanding the stages of grief and learning how to 'direct' people through laying their loved ones to rest. The industry really is about building relationships and knowing how to talk to people. Relationships with the community, building relationships with clergy at numerous churches, florists, different honor guard groups for veterans, writing obituaries, understanding how to submit obituaries to different newspapers, etc. Much like a wedding planner, a funeral director has to put all the pieces together to create and plan a funeral."

Meg: **"What made you pursue this field?"**

Cora Armstead: "Initially, when entering college, I wanted to become a veterinarian, so I pursued a bachelor of science degree in biochemistry.

During my freshman year, a close friend of mine passed away and I attended his funeral service. The condition and work that was done to him by the funeral home seemed to be subpar. Although I knew he was no longer alive, he just looked dead. Blue lips, sunken temples, blue fingernails . . . a scene from a horror movie. I didn't know anything about the funeral industry, but I knew that given the chance, I could do better. I felt as though my contribution to this world would be allowing every family I meet to say a proper goodbye to their loved one. I decided to switch majors and pursue mortuary science. I knew that I wanted to focus on embalming. Little did I know that there was so much more to funeral directing than the body preparation, and I fell in love with every aspect of the job."

Cemetery tours are popular in New Orleans, Louisiana where a high water table and muddy soil led to necessary aboveground crypts. The largest cemetery in the world is the Najaf cemetery in Iraq. More than five million people are buried there.[4]

Kelly: **"Is there anything else you would like readers to know about you or your field?"**

Cora Armstead: "The funeral industry isn't static, the trends are always changing. With final disposition of the deceased, most people think of either cremation or burial. Nonetheless, the options are endless. Cremation is normally done in a retort, using heat and fire. Another alternative for cremation is alkaline hydrolysis. Instead of heat and fire, this process uses an alkaline water solution to break down the body (just like fire cremation, what is left in the chamber after the alkaline hydrolysis is the skeletal remains). There is also natural burial where the decedent remains unembalmed and is buried in a shroud without any permanent marker (or tombstone). With natural burials, usually seeds are planted above the burial site. After cremation or alkaline hydrolysis, one can be buried, interred in a columbarium, or scattered in a meaningful location.

Many families also choose to take a portion of the cremated remains home. The family can place the cremated remains into jewelry that has a chamber for the ashes. There are companies that can turn portions of the ashes into diamonds or blow it into glass jewelry or statues. Some companies can place cremated remains into a firework or some families choose to have the ashes tattooed into their skin. There is also a company that can send cremated remains into space to orbit the atmosphere.

A new trend that has been occurring is that some families will hire professionals to come into the funeral home and carefully remove a tattoo (including the surrounding skin) to be preserved and displayed in a frame.

Another trend that has popped up in recent years is called 'extreme embalming.' Extreme embalming is when someone is embalmed in a way where they are staged during the funeral. This is popular in New Orleans. For example, there was a funeral where the deceased loved his motorcycle. The funeral home embalmed him in the position of riding his motorcycle, and at his funeral, he was staged riding his actual motorcycle. This practice is so new and uncommon, and it definitely can be off-putting compared to the traditional way of doing funerals. I find it incredibly fascinating; if families want to honor their loved one that way, I am in awe with the funeral directors who are skillful enough to make it happen.

My favorite way to memorialize someone with cremated remains is this company called The Living Urn. This company plants a tree mixed with cremated remains in honor of the family's loved one. They also have options for house plants such as bonsai trees and other plants. I love that as technology advances, so do the ways of how to meaningfully lay our deceased to rest."

We learned so much during our interview with Cora Armstead and are contemplating our own funeral plans. Readers, what will you do?

SECTION SEVEN
WITCHES IN LOVE

CHAPTER TWENTY-ONE
TEEN WITCH

The film *Teen Witch* (1989) centers on a girl and her obsession with her high school love interest. Having been a teen girl, I (Kelly), can absolutely empathize with Louise (Robin Lively) and her crush on Brad (Dan Gauthier) in the movie. Even the whole cheerleading squad proclaims "We Like Boys" with a choreographed musical number in the locker room! (Don't put it past me. I may have done something just as dramatic in my teen years . . . or more recently!) What is the science behind strong teen infatuations? Blame it on the hormones. Teenagers tend to be emotional, but throw in hormones? Everything gets elevated. Hormones are the chemicals responsible for sexual development and physical growth that can start as early as age seven or eight and carry through to adulthood. To begin, the brain releases gonadotropin-releasing hormone (GnRH) which triggers the pituitary gland to secrete follicle-stimulating hormone (FSH) and luteinizing hormone (LH) into the bloodstream. In girls, FSH and LH instruct the ovaries to begin producing estrogen and eggs. In boys, the same hormones tell the testes to begin producing testosterone and sperm.[1] Hormones affect

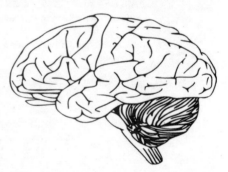

"The amygdala's connections with other parts of the brain are also strengthened during adolescence, and these changes appear to allow the adolescent brain to experience increasingly complex emotions."[2]

not only the body but also our emotions, moods, and sexual feelings.

Psychologists have identified three types of crushes: identity, romantic, and celebrity. Identity crushes have feelings of wanting to be like the person, trusting them as a leader, admiring them, and wanting to imitate them in some way. Romantic crushes involve the feelings of excitement and attraction. Celebrity crushes, as the name suggests, is a crush on a celebrity which can shape ideals and stir fantasies in the safety of never having in-person communication.[3] Adults should remember that crushes, no matter the type, are felt intensely as a teenager, so it's best not to make fun of them or take them too lightly.

All that being said, adults get crushes too. *Women's Health* magazine ran a piece that listed the reasons why having a crush as an adult isn't as exciting. There may be fewer options and you may think more logically about the chances with your crush, which could be less fun.[4] Like teenage crushes, adult crushes are usually built on fantasy. We tend to crush harder on people we know less about, but it can be stopped with "strong willpower from the frontal lobe, meditation, discipline, and practice."[5]

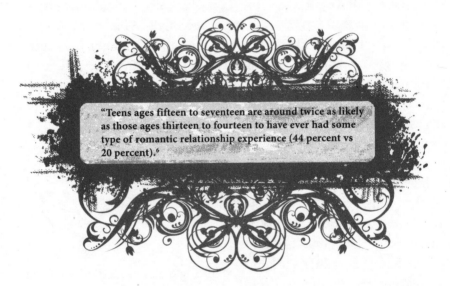

"Teens ages fifteen to seventeen are around twice as likely as those ages thirteen to fourteen to have ever had some type of romantic relationship experience (44 percent vs 20 percent).[6]

In the beginning of *Teen Witch*, Robin is shown dreaming. Right before she falls, she jolts awake. Why do our bodies do this? This is known as a hypnagogic jerk, named for the transition between sleeping and waking. Seventy percent of people experience this sensation, but it doesn't necessarily wake you up. There are theories as to why our bodies perform this involuntary action. One has an evolutionary basis. Primates' bodies may have tried to wake them up to readjust positions so they didn't fall out of a tree. Other theories include use of stimulants like caffeine, exercise too close to bedtime, or anxiety and stress.[7]

Teen Witch explores themes of popularity among teen girls, feelings of inadequacy, and discovering power. Louise changes into a more daring outfit at the school dance and gains confidence. Why do new outfits make us feel better? Scientists have concluded that clothes change the way we think and act and can affect others' perceptions of us. "Enclothed cognition" is the term given to how clothes shape our thoughts. There have even been studies to prove this:

> A study in *Social Psychological and Personality Science* found that those who wear formal business attire feel more powerful and in control of a situation, especially when going through a cognitive processing experiment. They also showed higher levels of intellect and creativity—even if they were on the same level as their casual-wearer counterparts. A study in the *Journal of Experimental Psychology* even found that wearing a suit gave you an edge in an argument over someone in sweats or casual clothing.[8]

Wear that outfit you love for a job interview or on a first date! It can only help your confidence.

The percentage of people who believe in reincarnation ranges from 12 percent to 44 percent depending on the country being surveyed. In the United States, it's 33 percent.[9]

Madame Serena (Zelda Rubinstein) reveals to Louise that they were both witches in a past life and were burned at the stake. In college, I was invited to a past life reading at a classmate's house. The woman in charge, before shaking my hand, said "Be careful of your thoughts now, I can read your mind!" She told me I don't like orange juice (she was right) and I should eat more kelp (I don't!). I didn't learn anything about my past lives that night, but it was an interesting experience.

According to the Pew Research Center, 33 percent of Americans believe in reincarnation, which is considered a "new age" belief.[10] Some children have claimed to have memories from past lives that have been checked and found to be factual. It usually occurs between the ages of two and five with memories fading around the age of seven. How do doctors explain this? Dr. Jim Tucker, director of the division of perceptual studies at the University of Virginia, believes these children's' stories and wants parents to know that this is not a sign of mental illness in the child. He recommends being open to what the child is saying and listening respectfully. Dr. Tucker encourages parents to avoid asking pointed questions but to write down information that the child shares in order to verify it later.[11]

Louise is told in *Teen Witch* that she'll come into her powers on her sixteenth birthday. This theme, of young women coming into their power during their teen years, has been explored in various films and television shows including *Carrie* (1976), *The Queen's Gambit* (2020), and *The Craft* (1996). Manifesting is bringing something into reality. What is the science behind this? The Law of Attraction is based on the belief that our thoughts can cause positive or negative occurrences in our lives. Using positive journaling or affirmations could then, in turn, manifest the goals you set for yourself. Scientifically, this could be related to the placebo effect (see page 20). If we think positive thoughts and believe in our goals, we will work toward them. If we have negative thoughts about a goal and are down on ourselves about it, we may self-sabotage and give up. The key is to be aware of our thoughts and the effect they have on our actions.

Louise uses a truth spell on her classmates in *Teen Witch*. Are there such things as truth serums? There are numerous drugs that have been used throughout history to try to extract information from people who are unwilling to provide it, but the ethics and legality of this practice is questionable. One issue that arises when using a drug to get someone to tell the truth is that they may become more susceptible to suggestion. Their memories may also not be as trustworthy compared to a sober mind. Various drugs have been used and tested as possible truth serums including LSD, marijuana, and scopolamine. All proved to make people more talkative but not necessarily truthful.

A 2016 study showed how dishonesty alters people's brains, making it easier to tell lies in the future. When people uttered a falsehood, the scientists noticed a burst of activity in their amygdala, but the more people were told to lie in the study, the less the amygdala fired.[12]

Like many teen coming-of-age stories, Louise loses her best friend as she gains popularity. How does fame or popularity affect our personalities and relationships? A study conducted over the course of ten years by the University of Virginia followed popular thirteen-year-olds to see how they developed after a decade. The results found that those kids, now twenty-three, were more likely to abuse alcohol and drugs, had a more difficult time making friends, and had less fulfilling romantic relationships than their less popular peers.[13] Another study of adults found that those who wished for close, personal relationships and helping others were happier and healthier than those who longed for fame or popularity.[14] At one point, Madame Serena tells Louise "The real magic is believing in yourself. If you can do that, you can make anything happen." Sounds like great advice for us all.

CHAPTER TWENTY-TWO
THE LOVE WITCH

"We are all born for love. It is the principle of existence, and its only end." This quote from Benjamin Disraeli could be the title character's mantra in *The Love Witch* (2016). The movie follows a witch, Elaine (Samantha Robinson), on a quest to find her "prince charming." The film is an homage to the era of the sixties and seventies and is visually stunning. Every detail, down to the costumes and set decoration, is curated to draw the audience into this quirky yet dark world. As one critic said, the title character is "a serial killer who thinks of herself as the star of a rom-com."[1]

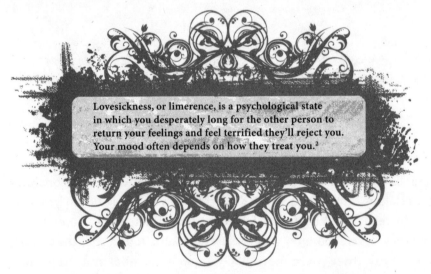

Lovesickness, or limerence, is a psychological state in which you desperately long for the other person to return your feelings and feel terrified they'll reject you. Your mood often depends on how they treat you.[2]

What is the science behind love? If you've ever been in love, you know the physical feeling of butterflies in your stomach and the constant, exciting thoughts about the person consuming your time. Falling in love affects us both physically and emotionally. "When in love, neurochemicals like dopamine and oxytocin flood our brains in areas associated with pleasure and rewards, producing physical and psychological responses like less perceived pain, an addictive dependence, and a stronger desire for sex with your partner."[3] Studies have shown that being in love can lower your blood pressure, help you to feel less stress, and feel less pain.[4]

Sociologist John Alan Lee suggests six different types of love. The first is eros, a passionate love based on attraction and romance. Physical

Studies have shown that viewing images of someone you find attractive can elicit a nonverbal response of pupil dilation.[5]

appearance tends to be important in this type of relationship and eros love can fade when looks do. Ludus is a more playful love that may not be long-lasting. Relationships based on Ludus may fizzle out fast if things get boring. Storge on the other hand is based on peace and slowness. This type of love we often find in friendships or friends who become a couple later. Pragma love is logical, based on a choice of commitment, not necessarily romance. Pragma couples may not have a lot of excitement in their relationship, but they are thoughtful. Mania, as the name implies, is manic. Couples who experience this type of love may make up and break up often. It's not a healthy type of love and can lead to jealousy and obsession. Last, agape love is selfless and unconditional. Although we may strive for agape love in our relationships, it's not always mutual.

In his book *The Five Love Languages* (1992), Gary Chapman outlines the five ways we express love and prefer to receive it. The first is words of affirmation. Some people need to hear positive things from their partner in order to feel loved and wanted. The second is quality time. There are those that need to spend time with their love interest to form a more meaningful connection. The third love language is receiving gifts. As the name suggests, gift giving can be an important part of a relationship. The fourth is acts of service for the other person and the fifth is physical touch. Certain people don't feel loved unless they have that component of physical touch and closeness with another person.[6]

Elaine claims she is addicted to love in *The Love Witch*. We all know certain people who have the constant need to be in a relationship. They're called serial monogamists and they jump from relationship to relationship, not staying single for very long. What do experts say about this? Sexologist Gigi Engle says if you're constantly looking for someone new to complete you, you may need to complete yourself first. "When you are regularly looking for someone new, you don't learn. You stay busy to avoid doing the work internally. You can't stay out of relationships long enough to learn from past mistakes. If you can't even bear to stand on your own two feet for any significant period of time, how can you expect to form a stable and equal partnership?"[7] She suggests staying single for a while after a breakup to get to know yourself better and not rush into a new relationship too quickly. Some people completely lose all sense of self in relationships and have molded their life to fit their partners. Taking this time to be single can help solidify who you are, your likes, your habits, and reconfirm your goals in life and a partner.

Can people be addicted to love? Sandor Rado popularized the term "love addict" in 1928 and described it as "a person whose needs for more love, more succor, more support, grow as rapidly as the frustrated people around her try to fill up what is, in effect, a terrible and unsatisfiable inner emptiness."[8] Although not listed in the Diagnostic and Statistical Manual of Mental Disorders, love addiction is classified similarly to those obsessions like sex or gambling. Love addicts get addicted to the feeling of being in love and it often stems from former feelings of abandonment, the need to constantly be around others, and not having healthy boundaries. Sex and Love Addicts Anonymous was formed in 1976 and created a twelve-step program, similar to Alcoholics Anonymous, to help people overcome their addiction to love. One of the main tenets is to identify bottom-line behaviors which are "any sexual or emotional act, no matter what its initial impulse may be, which leads to loss of control over rate, frequency, or duration of its occurrence or recurrence, resulting in spiritual, mental, physical, emotional, and moral destruction of oneself and others."[9]

Elaine in *The Love Witch* does not heed the advice of taking time to be alone after a breakup and casts a spell to find a new love. Are there such things as love potions? Although no elixir exists to make someone fall in love with you, there are pheromones that contribute to feelings of attraction. Whether it be a person's natural musk or a perfume or cologne, olfactory communication and the sense of smell play a part in how we view others. Aphrodisiacs are natural substances that can also

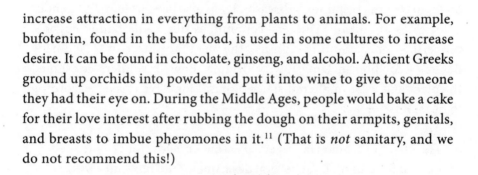

"Love can't give you the flu. But the hormone fluctuations associated with love and heartbreak—particularly the stress hormone cortisol—can prompt physical symptoms that affect your long-term health."[10]

increase attraction in everything from plants to animals. For example, bufotenin, found in the bufo toad, is used in some cultures to increase desire. It can be found in chocolate, ginseng, and alcohol. Ancient Greeks ground up orchids into powder and put it into wine to give to someone they had their eye on. During the Middle Ages, people would bake a cake for their love interest after rubbing the dough on their armpits, genitals, and breasts to imbue pheromones in it.[11] (That is *not* sanitary, and we do not recommend this!)

"Falling in love prompts your brain to ramp up production of certain hormones, including dopamine, oxytocin, and norepinephrine."[12]

Practicing witches have been casting love spells for centuries, but is this practice ethical? Writer Abby Lee Hood said, "More than one hundred practicing witches shared their thoughts in a 2019 survey I conducted about magick. Of the respondents, 63 percent said they had cast love spells in the past, but 87 percent said they no longer do. Many expressed misgivings about spells directed at individuals because they violate free will and consent but highlighted other types of more ethical love magick."[13] These include self-love spells or other ways to make yourself feel more attractive. *Making* someone fall in love with you, by any means, is a violation of consent.

Devil's weed contributes to a death in *The Love Witch*. How does this plant affect the human body? Devil's weed, also known as devil's snare or jimsonweed (see page 80), is a flowering plant in the nightshade family. It has been used throughout history in medicine from setting broken bones to treating respiratory diseases. It has also been used in Indigenous cultures as an aid in visionary ceremonies due to its hallucinogenic properties. One famous overdose of the plant took place in 1676 during Bacon's Rebellion in Jamestown, Virginia. English soldiers consumed the plant and were in altered states of mind for eleven days! According to observers:

The effect of which was a very pleasant comedy, for they turned natural fools upon it for several days: one would blow up a feather in the air; another would dart straws at it with much fury; and another, stark naked, was sitting up in a corner like a monkey, grinning and making mows [grimaces] at them; a fourth would fondly kiss and paw his companions, and sneer in their faces with a countenance more antic than any in a Dutch droll. In this frantic condition they were confined, lest they should, in their folly, destroy themselves—though it was observed that all their actions were full of innocence and good nature. Indeed, they were not very cleanly; for they would have wallowed in their own excrements if they had not been prevented. A thousand such simple tricks they played, and after eleven days returned themselves again, not remembering anything that had passed.[14]

Devil's weed is widely abused but addiction to it is rare due to its negative effects on the user. (People try it once and don't want to try it again.)

Anna Biller not only wrote and directed *The Love Witch* but she also edited it, designed the costumes, composed the music, and was the production designer. As film critic David Erhlich said:

> Biller shows an incredible command of tone and texture, the committed sensuality of her production design allowing her to thread the needle between camp and classicism. But the degree of consistency on display here is only possible because every department is on the same page, from the marvelously florid costumes (which Biller designed herself) to the performances of each supporting actor, all of whom is attuned to the weirdo vibes the movie is putting out there.[15]

We had the opportunity to interview Biller about the film.

Kelly: **"What was your first introduction to witches in media when you were young? How did that shape your viewpoint of them?"**
Anna Biller: "The TV show *Bewitched* (1964–1972), *The Wizard of Oz* (1939), *Bell, Book and Candle* (1958), and witches in fairy tales. I learned from *The Wizard of Oz* that there are good witches and bad witches, beautiful and ugly witches. In fairy tales, witches were usually evil women who wanted to do harm to female children or young women. *Bewitched* and *Bell, Book and Candle* taught me that a regular housewife can be a witch, and associated femininity—that is, just being a woman—with being a witch. So, all of the qualities that made witches magical, mysterious, clever, alluring, ingenious, glamorous, capable, scary, etc. were really just female qualities. This image of the glamorous modern witch was my favorite."

Meg: **"What research or aspects of witchcraft were important for you to portray accurately in *The Love Witch* and which ones were you more willing to be loose with?"**
Anna Biller: "I'm not interested in portraying anything inaccurately. I did loads of research, and I discovered that witchcraft as it's practiced

now is very young and very diverse. I saw mixed covens, for example, with satanists, Wiccans, druids, thelemites, people who practice Santeria or voodoo—whatever. Also, some people like costumes and elaborate altars, and some rituals are incredibly minimal. There are sexual groups, family-oriented groups. Going to rituals in Los Angeles reminded me of going to bohemian parties in Los Angeles—just a mix of artsy folks that I would already know or do already know. And in one class I went to, it wasn't artists and bohemians but more like businesspeople, really square people. People from all walks of life. My dad used to hang out at Samson DeBrier's house in Los Angeles, with Cameron and Kenneth Anger and all those Los Angeles witches. For a lot of 1960s witches, witchcraft seemed to be a lot about performance art and theater. For others, it was about free sex or exploring spirituality. My dad liked to read the books on Eastern religion and philosophy in DeBrier's library, some of them rare, banned books, and he met and became friends with Curtis Harrington there.

Growing up in LA, witches were just part of any scene you might belong to. And people believe all sorts of esoteric things. I don't really know how you can be loose with something that is already so loose. I tried to base my rituals on Alexandrian and Gardnerian rituals, and on Crowley, which seem to be the closest to a 'tradition' that witchcraft has, unless you go back to early pagan practices."

Kelly: "**You explore how men can't necessarily handle their emotions when hit with the love spell in the film. Do you think this reflects the 'Cowboy Syndrome' (males being shamed for and discouraged from sharing their emotions) in our society?**"

Anna Biller: "I think everyone is afraid to love or to open themselves up and be vulnerable to another person. I'm not an expert on nature vs nurture, but in my opinion the fear men have of falling in love with a woman is primal—the fear of being swallowed up and engulfed by the mother. And people *should* be cautious when giving someone their heart. I also think it's false to say that men don't share their emotions. Maybe not publicly, maybe not to other men, but they certainly share them with women. And straight women see it all—all the vulnerability, the pain, the grief, and especially the rage. And it's intense. The reason we don't see more men sobbing and screaming in films is that most men don't experience other men that way, or don't want to think of themselves in those moments, and most films are made by men. But we women see it because our men show us everything."

Meg: **"You express through dialogue in the film that the female form can be a powerful tool to use. How do you view the male gaze in films and, as a director, how did this impact how you shot the female form in *The Love Witch*?"**

Anna Biller: "I think it's great that we became aware of something called the male gaze in the 1970s, and that we can apply it critically now, but the discussion around the male gaze is incredibly lacking in nuance. For instance, it's universally agreed that Hitchcock used the male gaze in all of his shots, but that's really unhelpful to me as a way of understanding his films. I don't watch his films with a male gaze. I watch them with a female gaze. And it makes perfect sense to view them that way. So I would say his films contain a double gaze. The desiring look from the man onto the woman is matched by her acknowledging and responding to that gaze, and by gazing at him in turn. They're both sexualized, they're both looking. So why would I take his point of view? The fact is that I don't. Many films don't allow for that double gaze, but his films do.

To me, it's more helpful to think about why I might become alienated from a film. Usually, it's not the absence of women, or the sexualization of women, or men gazing at women that alienates me. It's more humiliation of women, women as grotesqueries, women only in a film to be fetishized while screaming or scrutinized while dead, stuff like that. It's not just my opinion that the female body is powerful. The female form *is* powerful because it exerts a magical control over men. Elaine shows men her body, and the men create the magic in their minds. The audience knows full well that she doesn't need drugs or spells to enchant them. Her power is the power of the femme fatale—in witchcraft parlance, 'sex magick.'"

Kelly: **"Tell us about your next projects!"**

Anna Biller: "I'm working on two films right now: *Bluebeard* and *The Face of Horror*. Bluebeard is a woman-in-peril film based on classic Hollywood women's pictures like *Rebecca* (1940) and *Gaslight* (1944), and takes place in a Gothic castle, although it is set in the present. *The Face of Horror* is based on a Japanese ghost story but is set in medieval England and is a gory revenge tale."

We can't wait to see Anna Biller's new projects and are thankful she took the time to talk with us!

Elaine is accused of driving a man to die by suicide in *The Love Witch*. Can people really be charged with this crime? It is true that this can happen. The manager of a Dairy Queen in Missouri was charged in the death of an employee after the way she treated him on the job, which led to his subsequent suicide. A nurse in Minnesota was charged after encouraging and informing people how to die by suicide online. According to a law firm, "In a personal injury lawsuit against someone who allegedly caused someone else's suicide, the main question is going to be whether or not the defendant's actions were significant enough to cause the person's suicide. Specifically, was it reasonably likely that the deceased person would have committed suicide as a result of the defendant's actions?"[16] Perhaps the most well-known and recent example of being charged in someone's death by suicide is the case of Michelle Carter and Conrad Roy. Carter had encouraged Roy to commit suicide, which he followed through with in 2014, and Carter was found guilty of involuntary manslaughter.

It's not revealed at the end of *The Love Witch* if Elaine will be charged in the stabbing death of Griff (Gian Keyes). Author Annette LePique wrote, "Elaine's killing of her cold-hearted lover, then, is a grim and bloody birth. This analogy makes sense—the murder marks the moment Elaine ceases to be an object of desire and becomes an active, desiring subject with no need for subterfuge or dishonesty . . . she no longer has to be perfect. She is not killed but kills. The cycle is broken."[17]

CHAPTER TWENTY-THREE
PENNY DREADFUL

The *Penny Dreadful* (2014–2016) TV series is an amalgamation of classic horror novels from classic Victorian literature including *Dracula* (1897), *Frankenstein* (1818), *The Picture of Dorian Gray* (1890), and *Strange Case of Dr. Jekyll and Mr. Hyde* (1886), as well as iconic figures from the time like Sherlock Holmes and Jack the Ripper. Setting all these characters in the same world makes for an epic and brilliant series. Created by John Logan, *Penny Dreadful* delves into everything from vampires to witches with a keen eye on modern tropes and serves as a love letter to the classic gothic genre. In reference to the season two episode "The Nightcomers," Logan said:

> The greatest horror in *Penny Dreadful* is the horror of people, the horror of the way we interact with one another, not the horror of a witch or a werewolf or a vampire. So I thought it was very important, going into our second season, to write an hour that really dramatizes that. It's something we would talk about. I talk about it in interviews. The actors talk about it, we talk about it on set. At heart, this is a story about people in pain, people searching and trying to find something. It's not about the tropes of Victorian horror. I wanted to write an episode that really dramatized that for the character.[1]

Set in 1891, we meet Vanessa (Eva Green), who we come to find out is a witch. Other witches are introduced later in the series and are referred to as Nightcomers. The powers they possess range from agility to eternal youth, shapeshifting to psychokinesis.

Vanessa is institutionalized and subsequently demoralized through many practices like getting sprayed with a hose, having her hair trimmed down to a buzz cut, and getting shock therapy. What is the history and psychology behind these things? Historically, women have been diagnosed with conditions such as hysteria when they were acting "out of order" in any sort of way. Many were institutionalized for disorders that medically did not exist, but psychiatrists were using these diagnoses to police women's behavior. Oftentimes psychiatrists were hired by husbands or fathers to investigate their wives' or daughters' "abnormal" behaviors. This then male-dominated field allowed thousands of women

First lady Mary Todd Lincoln was forcibly institutionalized during her life and her behavior was labeled as hysterical. A recent book suggests she may have had pernicious anemia due to low B12 levels which would explain her symptoms.[2]

to be committed to asylums for reasons ranging from masturbation to being overly educated. One woman, Elizabeth Packard, was locked up for disagreeing with her husband's extreme religious beliefs. After her release, she founded the Anti-Insane Asylum Society and campaigned for divorced women to retain custody of their children.

The Cut-Wife (Patti LuPone) uses totems for protection. How do various cultures view totems? They have been found in Indigenous cultures of the United States and are defined as "a spirit being, sacred object, or symbol of a tribe, clan, family, or individual. Some Native American tribes' tradition provides that each person is connected with nine different animals that will accompany him or her through life, acting as guides."[3]

Some examples of totems include the rabbit which represents being artistic, in motion, and joyful, the moose which represents wisdom and being headstrong, and the eagle which represents freedom and intelligence. This practice of totemism has been found across Indigenous cultures throughout the world. Totem poles of the Pacific

Examples of totem poles in Stanley Park, Vancouver.

Northwest of North America are monumental poles of heraldry that function as crests of families or chiefs. They tell stories or commemorate special occasions and are read from the bottom to the top. Other physical totems for protection may include objects with symbols on them, such as the lucky hand or the eye in the triangle, imbued with a spell.

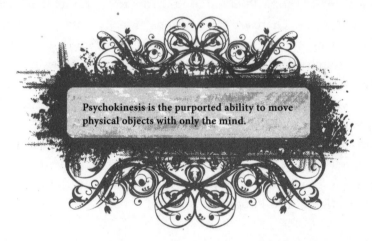

Psychokinesis is the purported ability to move physical objects with only the mind.

Vanessa and Frankenstein's monster (Rory Kinnear) talk of being in love and how to act. They recite poetry and practice dancing together. Be still my heart! What do relationship experts say is the best way to impress a first date? In scientific studies, "Openings involving jokes, empty compliments, and sexual references received poor ratings. Those revealing, e.g., helpfulness, generosity, athleticism, 'culture,' and wealth, were highly rated."[4] According to experts, the key to having a good conversation is reciprocity. Share similar stories, engage, ask questions, and listen. Maintaining enough eye contact and putting away distractions, like a cell phone, make a big difference in building rapport.

Penny Dreadful wasn't the first time Eva Green portrayed a witch. She also portrayed witches in *The Golden Compass* (2007) and *Dark Shadows* (2012).

The witches in *Penny Dreadful* use mirrors to transport themselves in and out of places. There are superstitions in many cultures regarding mirrors covering everything from mirrors stealing souls to creatures escaping through them. The urban legend of Bloody Mary has been passed down through generations and one variation reveals her to be a witch. Psychologists agree that these stories and urban legends play into our societal fears and insecurities about the unknown and can fuel our minds to scare ourselves. Directly related to the Bloody Mary ritual, experts have postulated for centuries the meanings behind mirrors and menstruation. From a study entitled "Bloody Mary in the Mirror: A Ritual Reflection of Pre-Pubescent Anxiety,"

A mirror is an obvious source of narcissistic pleasure (or concern) in this respect. Curiously, Aristotle is alleged to have said that "if a menstruating woman looks into a mirror, not only is the polish lost, but the person who next looks into the mirror will be bewitched. Pliny, speaking of this tarnishing effect on mirrors, says the polish can be restored by having the same woman look steadily upon the back of the mirror." Anthropologist Wallace reports that among the Mohave, a girl experiencing menses "must not look at her image in water or in a mirror or she will become cross-eyed." With regard to the supposed bewitching effect of menstruating women gazing into mirrors, we recall that the Knapps wondered why the comic strip figure of Mary Worth had become a mirror witch . . . . In a fascinating discussion of the folklore of menstruation, it has been suggested that "in folklore, the conclusion is that menstruation causes a woman to act like a witch" Devereux goes so far as to claim that 'The Menstruating Woman as a Witch" is the central theme of the psychoanalytic approach to menstruation.[5]

Vanessa is clearly depressed after Ethan (Josh Hartnett) leaves. What is the best way to get over a breakup? Experts say to remind yourself that these feelings won't last forever. The ennui felt post-split is temporary even though intensely felt. Surround yourself with nourishing people, not noxious ones. Noxious people will add fuel to the fire, so to speak, and won't be helpful in navigating your feelings. Nourishing people tend to be good listeners, will be honest with you, and will engage in meaningful, frequent interactions. In a 2018 study, participants were told four ways to get over an ex: think about the negative traits of the person, read and believe positive self-affirmations regarding their own feelings, find positive distracting things to think about, and not think about anything at all (a control condition). The findings of the study showed that people were able to change their feelings about their breakups, but it would take time and practice.[6]

It could also be said that Vanessa is suffering from a form of agoraphobia in the beginning of season three. What is the psychology behind this condition? Agoraphobia is an anxiety condition characterized by people not wanting to leave their home or not wanting to be in crowds. Symptoms can include increased heart rate and breathing, feeling faint or shaky, and feeling physical pain in the chest or stomach. Agoraphobia can be treated with a combination of psychotherapy and medication.[7]

Dr. Seward (Patti LuPone) uses hypnosis on Vanessa to help recall her memories, but hypnosis poses the risk of creating false memories as the result of suggestive, leading questions. Hypnosis has proven to be useful for burn victims but not any more effective than other tools for those who want to quit smoking.[8] Read more about hypnosis in our book *The Science of Monsters* (2019).

The series ended on a sad note (no spoilers), leaving fans feeling frustrated and disappointed. Series creator John Logan said, "Growing up as a gay man, before it was as socially acceptable as it is now, I knew what it was to feel different, to feel alienated, to feel not like everyone else. But the very same thing that made me monstrous to some people also empowered me and made me who I was."[9] Whether it was through the empathy building scenes of Frankenstein's monster at the theatre, Vanessa finding love and companionship for a brief time, or discovering the brilliant and beautiful backstories of other characters, *Penny Dreadful* will live on in my memory as one of my favorites.

CONCLUSION

Through the experience of writing this book, we both came to appreciate the multi-faceted nature of the witch. A predominantly female archetype, the witch cannot be reduced to just a horror film staple. She is an amalgam of all the witches who have come before and after. She is the spirited face of nonconformity. She revels in her powers, whether they are supernatural or grounded in nature itself. Never will we hear the term "witch" without thinking of the incredible interviewees for this book, as well as the fictional witches who have graced our lives. May every reader embrace their inner witch. For she is a special kind of magic. She is inside you, waiting to be unleashed.

ABOUT THE AUTHORS

Kelly Florence teaches communication at Lake Superior College in Duluth, Minnesota, and is the creator of the *Be a Better Communicator* podcast. She received her BA in theater from the University of Minnesota-Duluth and her MA in communicating arts from the University of Wisconsin-Superior. She has written, directed, produced, choreographed, and stage managed for dozens of productions in Minnesota including *Carrie: The Musical* through Rubber Chicken Theatre and *Treasure Island* for Wise Fool Theater. She is passionate about female representation in all media and particularly the horror genre.

Horror and suspense author **Meg Hafdahl** is the creator of numerous stories and books. Her fiction has appeared in anthologies such as *Eve's Requiem: Tales of Women, Mystery, and Horror* and *Eclectically Criminal.* Her work has been produced for audio by *The Wicked Library* and *The Lift,* and she is the author of three popular short story collections including *Twisted Reveries: Thirteen Tales of the Macabre.* Meg is also the author of the novels *This World is Nothing but Evil, The Darkest Hunger, Daughters of Darkness,* and *Her Dark Inheritance* called "an intricate tale of betrayal, murder, and small town intrigue" by *Horror Addicts* and "every bit as page turning as any King novel" by *Rochester Women Magazine.* Meg lives in the snowy bluffs of Minnesota.

Together, Kelly and Meg have written five books: *The Science of Monsters, The Science of Women in Horror, The Science of Stephen King, The Science of Serial Killers,* and now *The Science of Witchcraft.* They cohost the *Horror Rewind* podcast and write and produce horror projects together.

ACKNOWLEDGMENTS

Thank you to Nicole and everyone at Skyhorse!

Thank you to Josh for believing in us and challenging us in all that we do.

Thank you, Kelly and Madeline, for inspiring us and supporting our vision.

Thank you to Gail and Nancy for championing us on this journey!

Thank you to Ben, Cora, Nicole, Brett, Drew, Anna, Linda, Axelle, Peaches, Jinkx, Stephanie, and Payton for your time and fascinating interviews!

Thank you to our families for their constant love and support.

Thanks to Karmen, Stacey, and Sam for looking out for our best interests.

And, as always, thank you to our Rewinders for your unrelenting support . . . we'll see you in the horror section!

ENDNOTES

Chapter One: The Wizard of Oz

1. Bailey, M. D. (2010) *Battling Demons: Witchcraft, Heresy, and Reform in the Late Middle Ages.* Pennsylvania State University Press.
2. Bergman, Jess. (October 30, 2015) "A Literary History of Witches." *Literary Hub.*
3. Davies, Owen. (2009) *Grimoires: A History of Magic Books.* Oxford University Press.
4. Gourley, Catherine. (1999) *Media Wizards: A Behind-the-Scene Look at Media Manipulations.* Brookfield, CT.
5. Culver, Stuart. (1988) "What Manikins Want: The Wonderful Wizard of Oz and the Art of Decorating Dry Goods Windows." *Representations.*
6. (October 29, 2011) "A Scarecrow's Origin May Surprise You." *Jamestown Sun.*
7. Kurtz, Holly. (1999) "Harry Potter Expelled From School." *Denver Rocky Mountain News.*
8. Schwartz, Dan. (May 6, 2014) "The Books Have Been Burning." *CBC News.*
9. Tucker, Trisha. (June 25, 2017) "What do protests about Harry Potter books teach us?" *The Conversation.*
10. Wigington, Patti. (August 31, 2018) "Scarecrow History and Folklore." *Learn Religions.*
11. (May 13, 2016.) "*AFI's 100 Years . . . 100 Movies.*" AFI.com.
12. Bondarenko, Veronika. (August 22, 2019) "The Curse of Playing the Wicked Witch of the West." *Narratively.com.*
13. Ghosh, Sudip Kumar. (September 1, 2019) "The color of skin: green diseases of the skin, nails, and mucosa." *Clinics in Dermatology Journal.*
14. Guiley, Rosemary. (2010) *The Encyclopedia of Witches, Witchcraft and Wicca.*
15. Bell, Vaughan. (September 18, 2010) "The Strange-Face-in-the-Mirror Illusion." *Mind Hacks.*
16. Madrigal, Alexis C. (March 19, 2013) "Powerful Tornadoes Can Transport Photographs Over 200 Miles." *The Atlantic.*
17. Samenow, Jason. (January 24, 2017) "A Woman Flew Through a Tornado in a Bathtub and Survived." *The Washington Post.*
18. (April 13, 2020) "Tornado Blows Entire House Into Middle of Road." *WSB-TV.*
19. Lada, Brian. (2021) "The Science Behind How Tornadoes Form." *Accuweather.*
20. Baum, L. Frank. (1900) *The Wonderful Wizard of Oz.* George M. Jill Company.
21. Schwarcz, Joe, PhD. (March 20, 2017) "Would Dorothy and the Cowardly Lion Have Passed an Emerald City Entrance Drug Test?" *McGill.*
22. Gans, Andrew. (December 19, 2003) "DIVA TALK: A Chat with *Wicked*'s Idina Menzel Plus News of Cook and Ripley." *Playbill.*

23. Frenschkowski, Marco. (December 20, 2019) "Witchcraft, Magic and Demonology in the Bible, Ancient Judaism and Earliest Christianity," in *The Routledge History of Witchcraft*, ed. Johannes Dillinger Abingdon: Routledge.

Chapter Two: Sleepy Hollow
1. Katzman, Rebecca. (August 11, 2014) "Here's Why a Chicken Can Live Without Its Head." *Modern Farmer.*
2. (2021) "Mike the Headless Chicken Festival." Mike the Headless Chicken.org.
3. Puiu, Tibi. (October 8, 2013) "How Long Can a Person Remain Conscious After Being Decapitated?" *ZME Science.*
4. Byard, R. W., Gilbert, J. D. (2004) "Characteristic features of deaths due to decapitation." *Am J Forensic Med Pathol.*
5. Cassata, Cathy. (September 24, 2018) "Here's Why Being Scared Can Make You Faint." *Healthline.*
6. Upton, Roy. (March 2013) "Stinging nettles leaf (*Urtica dioica* L.): Extraordinary vegetable medicine." *Journal of Herbal Medicine.*

Chapter Three: Season of the Witch
1. Williams, Tony. (2015) *The Cinema of George A. Romero: Knight of the Living Dead.* Columbia University Press.
2. Warren, Derek. (March 14, 2018) "Cult Corner: Season of the Witch is an Interesting Departure for Horror Icon George Romero." *Nepa Scene.*
3. Daniel, Alex. (March 31, 2021) "Secret Meaning of Sixty Common Dreams, According to Experts." *Best Life.*
4. (2018) "That Age Old Question: How Attitudes to Aging Affect Our Wellbeing." *Royal Society for Public Health.*
5. Bryce, Amber. (March 15, 2020) "Gerascophobia, a Real Fear of Getting Older, is On The Rise." *Refinery 29.*
6. (April 14, 2020) "Empty Nest Syndrome." *Mayo Clinic.org.*
7. Sathyanarayana Rao, T. S. (October–December 2009) "The Biochemistry of Belief." *Indian Journal of Psychiatry.*
8. Kirsch I, Sapirstein G. (1998) "Listening to Prozac but Hearing Placebo: A Meta-Analysis of Antidepressant Medication." *Prevention & Treatment.*
9. Pietrangelo, Ann. (April 27, 2021) "The Effects of Cannabis on Your Body." *Healthline.*
10. (2021) "The Truth About Runaways." *The Polly Klaas Foundation.*
11. (2013) "Diagnostic Criteria 313.81 (F91.3)." *Diagnostic and Statistical Manual of Mental Disorders* (Fifth ed.). American Psychiatric Association.
12. Witrogen, Beth. (December 31, 2020) "Older Women, Younger Men." *Health Day.*
13. McCann, Ruairi. (2014) "All of Them Witches: Season of the Witch." *Berlin Film Journal.*
14. (October 2020) "2020 Accidental Gun Death Statistics in the US." *Aftermath.*

Chapter Four: We Have Always Lived in the Castle
1. Jackson, Shirley. (1962) *We Have Always Lived in the Castle*. Viking.
2. Franklin, Ruth. (August 15, 2016) "Tarot Cards." *Tinyletter.com*.
3. Jackson, Shirley. (1956) *The Witchcraft of Salem Village*. Random House: New York.
4. (June 21, 2021) "Bridget Bishop." *Salem Witch Museum*.
5. Hill, Frances. (1995) *A Delusion of Satan: The Full Story of the Salem Witch Trials*. Doubleday: New York.
6. Schiff, Stacy. (November 2015) "Unraveling the Many Mysteries of Tituba, the Star Witness of the Salem Witch Trials." *Smithsonian Magazine*.
7. Andrews, Evan. (March 18, 2014) "Bizarre Witch Trial Tests." *History.com*.
8. Walker, Rachel. (2021) "Cotton Mather." *University of Virginia*.

Chapter Five: Black Sunday
1. (2021) "The Nuremberg Virgin." *The Torture Museum*.
2. Goran, David. (November 2, 2016) "The Infamous Iron Maiden: A horrific form of execution intended to inspire terror." *The Vintage News*.
3. (February 19, 2019) "9 Brutal Torture Methods and Cruel Punishments." *The Scotsman*.
4. Pappas, Stephanie. (September 6, 2016) "Are Iron Maidens Really Torture Devices?" *Live Science*.
5. Konieczny, Peter. (August 31, 2021) "Why Medieval Torture Devices Are Not Medieval." *Medaevalists.net*.
6. (June 7, 2019) "Our Legal Heritage: Helen Duncan, Our Wartime 'Witch.'" *Scottish Legal News*.
7. (August 31, 2021) "The Last Witchcraft Trial." *Backpackervers.com*.
8. Chambers, Vanessa. (January 23, 2007) "The Witchcraft Act Wasn't About Women on Brooms." *The Guardian*.
9. Osbourne, Hannah. (August 2, 2019) "Face of 18th Century 'Vampire Witch' Who Was Tortured to Death to Be Revealed by Scientists." *Newsweek*.
10. Flynn, Daniel. (March 12, 2009) "'Vampire' Unearthed in Venice Plague Grave." *Reuters*.
11. Flynn, Daniel. (March 12, 2009) "'Vampire' Unearthed in Venice Plague Grave." *Reuters*.
12. Solly, Meilan. (October 15, 2018) "This Ancient 10-Year-Old Received a 'Vampire Burial' to Prevent Return From the Dead." *Smithsonian Magazine*.

Chapter Six: The Manor
1. Gellar, P. (April 8, 2017) "Dryad." *Mythology.net*.
2. Riggs, Ransom. (April 1, 2011) "Madagascar's Legendary Man-Eating Tree." *Mental Floss*.
3. Belew, Kate. (2021) "Maiden, Mother, and Crone: The Triple Goddess." *Magick and Alchemy.com*.
4. O'Connell, Jennifer. (October 28, 2017) "Witchipedia: Ireland's Most Famous Witches." *Irish Times*.

5. (2021) "Mark Steger." *IMDB.com.*

Chapter Seven: Eve's Bayou
1. (August 28, 2021) "Guide to Southern Gothic." 10 Dark Must Reads." *Reedsy.com.*
2. (October 28th, 2021) "Kasi Lemmons." *IMDB.com.*
3. (November 2, 2021) "What's Up with Dreams That Seem to Predict the Future?" *Healthline.*
4. Wiseman, Richard. (2011) "To Sleep, Perchance to Dream. How the Science of Sleep Explains 'Precognitive' Dreams." *Skeptic Magazine.*
5. Cherry, Kendra. (July 30, 2021) "How Confirmation Bias Works." *Very Well Mind.com.*
6. Love, Daniel. (2021) "Happy Hatsuyume! The First Dream of the New Year." *World of Lucid Dreaming.com.*
7. Blagrove, Mark, French, Christopher C., and Jones, Gareth. (2006) "Probabilistic Reasoning, Affirmative Bias and Belief in Precognitive Dreams." *Applied Cognitive Psychology.*
8. Mikanowski, Jacob. (September 28, 2016) "How to Read the Bones Like a Scapulimancer." *JStor Daily.*
9. Wigington, Patti. (September 25, 2019) "Bone Divination." *Learn Religions.com.*
10. (November 3rd, 2021) "Devil's Shoe String." *The Witch Depot.*
11. (November 3rd, 2021) "Hoodoo vs. Voodoo." *Difference Between.net.*

Chapter Eight: Gretel and Hansel
1. Zipes, Jack. (2013) "Abandoned Children ATU 327A—Hansel and Gretel." *The Golden Age of Folk and Fairy Tales: From the Brothers Grimm to Andrew Lang.* Hackett Publishing.
2. Collis, Clark. (August 26, 2019) "'It' star Sophia Lillis goes to a dark place in first look at Osgood Perkins's 'Gretel & Hansel.'" *Entertainment Weekly.*
3. Luu, Chi. (May 2, 2018) "Fairytale Language of the Brothers Grimm." *JSTOR Daily.*
4. Roffman, Michael. (January 30, 2020) "Gretel and Hansel score heading to vinyl, hear exclusive track "By The River": Stream". *Consequence of Sound.*
5. Beyer, Catherine. (July 8, 2019) "Geometric Shapes and Their Symbolic Meanings." *Learn Religions.*
6. Briggane, Eric. (April 29, 2014) "How to Tell the Difference Between Poisonous and Edible Mushrooms." *Wild Food UK.*
7. Smith, Melinda et al. (September 2020) "Emotional Eating and How to Stop It." *HelpGuide.org.*
8. Stockton, Nick. (October 28, 2014) "What's Up With That: People Feel Weather in Their Bones." *Wired.*
9. Glaveski, Steve. (August 28, 2018) "The Inverted U: Why Getting Too Much of a Good Thing Can Be Really Bad For You." *Noteworthy.*
10. (2021) "Five People Who Ate Themselves to Death." *Tastemade.*
11. Mayor, Adrienne. (2009) *The Poison King: The Life and Legend of Mithradates, Rome's Deadliest Enemy.* Princeton University Press.

12. Osternath, Brigitte. (December 1, 2018) "German patient is immune to highly poisonous ricin." *DW.com.*
13. Myers, Mara. (November 14, 2017) "Fear, Bravery, and Courage." *Medium.*

Chapter Nine: Maleficent
1. Sivan. (2014) "The Secret History of Maleficent: Murder, Rape, and Woman-Hating in Sleeping Beauty." *Revisiting History.*
2. Bates, David. (2001) "The Prognosis of Medical Coma." *Journal of Neurology, Neurosurgery & Psychiatry.*
3. Russon, Mary-Ann. (June 24, 2014) "International Fairy Day: Where Did the Fairies Come From?" *International Business Times.*
4. Bader, Christopher D., F. Carson Mencken, and Joseph O. Baker. (2010) *Paranormal America: Ghost Encounters, UFO Sightings, Bigfoot Hunts, and Other Curiosities in Religion and Culture.* New York; London: NYU Press.
5. Robertson, Colin. (March 11, 2016) "The Science Behind Guilt and Why It Is Never the Answer." *Medium.*
6. Wishhover, Cheryl. (May 19, 2014) "What it Took to Turn Angelina Jolie into Maleficent." *Fashionista.*
7. Haseltine, Eric. (April 25, 2015) "Yes, You Have a Sixth Sense, and You Should Trust It." *Psychology Today.*
8. Cristol, Daniel A. (May 1, 1999) "Avian prey-dropping behavior. II. American crows and walnuts." *Behavioral Ecology.*

Chapter Ten: The Wretched
1. (2021) "Saturn Devouring His Son." *Visual Arts Cork.com.*
2. Kroll, David. (October 31, 2017) "The Origin Of Witches Riding Broomsticks: Drugs From Nature, Plus Shakespeare." *Forbes.*
3. Vickery, Roy. (1983) "Lemna Minor and Jenny Greenteeth." *Folklore.*
4. Collins, Siobain. (March 14, 2019) "The Cailleach: Irish Myth, Legend and the Divine Feminine." *Folklore Thursday.*
5. (June 14, 2019) "Hudson Man Pours Salt on Feet to 'Ward Off Evil Spirits' at Walmart." *WIVB TV.*

Chapter Eleven: The Autopsy of Jane Doe
1. Soniak, Matt. (February 15, 2012) "Why are Unidentified People Called John or Jane Doe?" *Mental Floss.*
2. Dr. Munroe, Ranald and Dr. Munroe, Helen M.C. (2008) "Estimation of Time Since Death." *Animal Abuse and Unlawful Killing.*
3. Kanchan, Tanuj, Krishan, Kewal and Shrestha, Rijen. (April 20, 2021) "Methods Of Estimation Of Time Since Death." *Statpearls.*
4. Carlton, Genevieve. (October 4, 2019) "Even In Enlightenment Britain, You Could Be Sentenced To Death By Being Crushed With Huge Weights." *Ranker.*
5. Bachman, Mara. (April 14, 2020) "The Autopsy of Jane Doe: The Witch and Powers Explained." *Screenrant.*

6. Snyder, Heather, and Giles Corey. (2001) "Salem Witch Trials Documentary Archive and Transcription Project." *Virginia.edu.*
7. Ellison, George. (July 15, 2020) "Jimson Weed has a Long and Lethal History." *Smoky Mountain News.*
8. (October 2008) "Learning From the Dead." *Popular Science.*
9. (2021) "History of the Autopsy." *Mopec.com.*
10. (January 22, 2010) "Egyptian Woman's Death Resolved?" *Current Science.*
11. Collis, Clark. (September 27, 2016) "Autopsy of Jane Doe: How an Actress Played Dead for Horror Film." *EW Magazine.*
12. Reesman, Bryan. (February 4, 2020) "Geek School: The Art of Playing Dead." *Syfy Wire.*
13. Freedman, David. H. (August 30, 2012) "20 Things You Didn't Know About Autopsies." *Discover Magazine.com.*

Chapter Twelve: Don't Knock Twice

1. Clifford, Marissa. (November 3, 2017) "The Enduring Allure of Baba Yaga, an Ancient Swamp Witch who Loves to Eat People." *Vice.*
2. Holub, Christian. (April 15th, 2019) "How the New Hellboy Movie Brought Baba Yaga to Life." *Entertainment Weekly.*
3. H, Jim. (2021) "Baba Yaga: The Scary Witch of Slavic Folklore." *Historic Mysteries. com.*
4. Armknecht, Megan, Forrester, Sibelan and Rudy, Jill Terry. (2017) "Identifying Impressions of Baba Yaga: Navigating the Uses of Attachment and Wonder on Soviet and American Television" *Marvels & Tales.*
5. Hammer, Heather. (October 2002) "Runaway/Thrownaway Children: National Estimates and Statistics." *Office of Juvenile Justice and Delinquency Prevention.*
6. Hargitai, Quinn. (February 19, 2018) "The Strange Power of the Evil Eye." *BBC. com*
7. Davis, Lisa. (November 26, 2019) "Why Youth Run Away: National Runaway Prevention Month." *Family Resources Inc.org.*
8. Kane, Jenny. (July 10, 2019) "Baba Yaga's House moving to Burning Man's Fly Ranch after 2018 debut in Black Rock City" *Reno Gazette Journal.*
9. Huffman, Carla. (December 8, 2020) "Passageways, Gates, Portals." *Myths, Symbols, and Play.*

Chapter Thirteen: The Conjuring

1. (July 20, 2021) "Ed & Lorraine Warren." *New England Society for Psychic Research.*
2. (August 2, 2021) "11 Things You Need to Know About Legendary Paranormal Investigators Ed and Lorraine Warren." *The Travel Channel.*
3. Guglielmi, Jodi. (June 10, 2016) "The Enfield Poltergeist: Inside the Real Story that Inspired The Conjuring 2." *People.*
4. Maggi, Armando. (2014) "Christian Demonology in Contemporary American Popular Culture." *Social Research: An International Quarterly.*
5. Alexander, Bryan. (July 22, 2013) "The True Story Behind the Conjuring." *USA Today.*

6. Tomaiolo, Kristen. (October 18, 2013) "Andrea Perron Returns to RI to Tell the True Story Behind 'The Conjuring.'" *The Independent.*
7. G, Danka. (April 11, 2021) "Italian-American Stregheria." *The Proud Italian.*
8. (July 26, 2021) "The Conjuring." *History V Hollywood.*
9. Boehm, Omri. (April 1, 2004) "Child Sacrifice, Ethical Responsibility, and the Existence of the People of Israel." *Vetus Testamentum.*
10. (August 2, 2021) "Eusebius of Caesarea." *Christianity Today.*
11. Yuhas, Alan. (March 31, 2021) "It's Time to Revisit the Satanic Panic." *The New York Times.*
12. Ajimbo, Doreen. (September 26, 2017) "Witch doctors sacrificing children in this drought-stricken African country." *USA Today.*
13. Linder, Douglas O. (2021) "The West Memphis Three Trials: An Account." *Famous Trials.*

Chapter Fourteen: The Craft

1. Bricker, Tierney. (May 3, 2021) "25 Bewitching Secrets About The Craft Revealed." *Eonline.*
2. Bastien, Angelica Jade. (October 27, 2017) "The Profound, Enduring Legacy of The Craft." *Vulture.*
3. Wigington, Patti. (November 27, 2019) "Gods and Goddesses of Death in the Underworld." *Learn Religions.com.*
4. Matteoni, Francesca. (2009) "Blood Beliefs in Early Modern Europe." *University of Hertfordshire Research Archive.*
5. Farzan, Noori Antonia. (September 3, 2019) "A Catholic school removed Harry Potter books from its library, warning that readers 'risk conjuring evil spirits.'" *The Washington Post.*
6. Beebe, Jessica. (January 10, 2021) "Gretel & Hansel: Every Horror Movie Based On The Brothers Grimm Story." *Screen Rant.*

Chapter Fifteen: Suspiria

1. Fisher, Russ. (October 26, 2015) "Black Gloves and Knives: 12 Essential Italian Giallo." *Indie Wire.*
2. De Quincey, Thomas. (1845) "Leavana and Our Ladies of Sorrow." *Blackwood's Magazine.*
3. (April 6, 2018) "Samhain." *History.com.*
4. Cooper, Calum. (October 9, 2020) "Suspiria." *In Their Own League.*
5. Lothspeich, Jessica. (April 3, 2019) "Heaven's Gate Mass Suicide in San Diego made World-Wide Headlines 22 Years Ago." *CBS 8 News.*
6. Chamberlain, Lisa. (2014) *Wicca Elemental Magic: A Guide to the Elements, Witchcraft and Magic Spells.* Chamberlain Publications.
7. (April 26, 2021) "Autumnal Equinox." *Britannica.com.*
8. (May 28, 2021) "Samhain Symbol." *Symbols Archive.*
9. Hogenboom, Melissa. (February 26, 2015) "Do Whales Have Graveyards Where They Prefer to Die?" *BBC.com.*
10. (April 26, 2021) "Cult Suicide." *Psychology Wiki.*

Chapter Sixteen: The Witches of Eastwick
1. Baym, Nina. (1984) "The Review of the Witches of Eastwick by Nina Baym." *Iowa Review*. Volume 14, Issue 3, Article 57.
2. (June 24, 2021) "Polyamory." *Psychology Today*.
3. Tymn, Michael E. (2009) "An 'Interview' with French Educator, Scientist, and Philosopher Allen Kardec." *White Crow Books.com*.
4. Martin, Juan. (October 25, 2018) "Gerascophobia, or the Fear of Aging." *CENIE. eu*.
5. Slonim, Jeffery. (December 13, 2011) "Matt Damon 'Cried Like a Baby' Around Snakes on Set." *People*.
6. Brewer, Geoffrey. (March 19, 2001) "Snakes Top List of American's Fears." *Gallup.com*.
7. White, Tracie. (February 1, 2006) "People Who Fear Pain are More Likely to Suffer It." *Stanford Report*.
8. Radford, Benjamin. (October 29, 2013) "Voodoo, Facts About Misunderstood Religion." *Live Science*.
9. (November 11, 2018) "Magical Dolls: A History of Poppets and Effigy Spells." *A Bad Witch's Blog*.
10. (May 26, 2020) "What You Need to Know about Poppet Magic Before You Practice." *Green Witch Farm*.
11. Burns, Janet. (August 6, 2015) "A Brief, Sticky History of Tarring and Feathering." *Mental Floss*.

Chapter Seventeen: Death Becomes Her
1. Watkins, Gwynne. (July 31, 2017) "'Death Becomes Her' Turns 25: Behind the Scenes of Its Oscar-Winning Special Effects." *Yahoo Entertainment*.
2. Stefansson, Halldor. (July 2005) "The Science of Ageing and Anti-Ageing." *EMBO Reports*.
3. (2021) "Aesthetic Plastic Surgery National Data Bank Statistics for 2020" *Surgery. org*.
4. (September 9, 2021) "History of Medicine: Ancient Indian Nose Jobs & the Origins of Plastic Surgery." *Columbia Surgery.org*.
5. Kita, Natalie. (May 16, 2020) "The History of Plastic Surgery." *Very Well Health*.
6. Miska, Brad. (July 10, 2012) "1992 Academy Award Winning Effects Work From 'Death Becomes Her'!" *Bloody Disgusting*.

Chapter Eighteen: A Discovery of Witches
1. Salzman Boston, Michelle. (May 7, 2013) "The Dark of Harkness." *USC Dornsife*.
2. Salzman Boston, Michelle. (May 7, 2013) "The Dark of Harkness." *USC Dornsife*.
3. Jones, Jane. (2021) "41 Dark Facts About Medieval Alchemy." *Factinate*.
4. Jones, Jane. (2021) "41 Dark Facts About Medieval Alchemy." *Factinate*.
5. Brenner, Manuel. (June 10, 2020) "Alchemy and the Problems of Modern Science." *Medium*.
6. Lyons, Martyn. (2011) *Books: A Living History*. California: J. Paul Getty Museum.

7. Taylor, Astrea. (June 24, 2019) "Intuitive Witch Walks: Your Passport To Another World." *Patheos.*
8. Wigington, Patti. (August 26, 2020) "How to Write Your Own Spell in 5 Steps." *Learn Religions.*
9. White, Adrian. (June 22, 2020) "11 Benefits of Burning Sage, How to Get Started, and More." *Healthline.*
10. Guzmán Gutiérrez, S. Laura, Reyes Chilpa, Ricardo, & Bonilla Jaime, Herlinda. (2014). "Medicinal plants for the treatment of "nervios", anxiety, and depression in Mexican Traditional Medicine." *Revista Brasileira de Farmacognosia.*
11. Stein, Gordon. (1996) *Encyclopedia of the Paranormal.* Prometheus.
12. (2021) "Strength." *The Tarot Guide.com.*
13. Newcombe, Rachel. (2021) "Pendulum Dowsing—An Introduction to Using a Pendulum." *Holistic Shop.*
14. Durn, Sarah. (January 17, 2019) "A Medievalist's Guide to Magic and Alchemy in *A Discovery of Witches.*" *Gizmodo.*

Chapter Nineteen: Buffy the Vampire Slayer

1. Mason, Jessica. (March 18, 2020) "Why Buffy's Willow Was Never My Witch." *The Mary Sue.*
2. Fearnow, Benjamin. (November 18, 2018) "Number of Witches Rises Dramatically Across U.S. as Millenials Reject Christianity." *Newsweek.*
3. Ray, Mark Rinaldi (2016-05-16). "As an opera, 'The Shining' knows it's good to be King." *Denver Post.*
4. Vukanović, T. P. (1957–1959) "The Vampire." *Journal of the Gypsy Lore Society.*
5. Hoffman, Leslie A.; Vilensky, Joel A. (August 1, 2017). "Encephalitis lethargica: 100 years after the epidemic." *Brain.*
6. Zorumski, Charles F. (October 26, 2018) "What Causes Alcohol-Induced Blackouts?" *Scientific American.*
7. Bartholomew, Robert; Wessely, Simon (2002). "Protean nature of mass sociogenic illness. *The British Journal of Psychiatry.*
8. Small G. W. (2002) "What we need to know about age related memory loss." *BMJ.*

Chapter Twenty: Chilling Adventures of Sabrina

1. Dasgupta, Shreya. (February 16, 2015) "Can Any Animals Talk and Use Language Like Humans?" *BBC.*
2. Sedgwick, Icy. (July 27, 2019) "Witches' Familiars: The Good, the Bad, and the Weird." *Icy Sedgwick.com.*
3. (2020) "NFDA 2020 Cremation and Burial Report." *National Funeral Directors Association.*
4. (July 12, 2016) "Najaf, Iraq: The world's biggest cemetery." *BBC News.*

Chapter Twenty-One: Teen Witch

1. (December 15, 2012) "Teenage Hormones and Sexuality." *Newport Academy.*
2. Davis, Matthew. (December 9, 2015) "Hormones and the Adolescent Brain." *Brain Facts.org.*

3. Pickhardt, Carl E. PhD. (September 10, 2012) "Adolescence and the Teenage Crush." *Psychology Today*.
4. Georgopulos, Stephanie. (April 19, 2016) "8 Reasons Why Being a Grown-Ass Woman With a Crush Totally Sucks." *Women's Health Magazine*.
5. Naftulin, Julia. (December 4, 2018) "This is Why You Develop Crushes, According to Science." *Insider*.
6. Lenhart, Amanda et al. (October 1, 2015) "Basics of Teen Romantic Relationships." *Pew Research Center*.
7. Holland, Kimberly. (November 12, 2019) "Twitching Before Falling Asleep: What Causes Hypnic Jerks?" *Healthline*.
8. Lopez, Carmen. (2021) "Look Good, Feel Great: The Psychology of Clothing." *Craving Current*.
9. Gecewicz, Claire. (October 1, 2018) "'New Age' beliefs common among both religious and nonreligious Americans." *Pew Research Center*.
10. Gecewicz, Claire. (October 1, 2018) "'New Age' beliefs common among both religious and nonreligious Americans." *Pew Research Center*.
11. (2021) "Advice to Parents of Children who are Spontaneously Recalling Past Life Memories." *University of Virginia*.
12. Stillman, Jessica. (October 2, 2018) "The Science of Lying; the More You Do It the Easier It Gets." *INC.com*.
13. Allen, Joseph P. et al. (2014) "What Ever Happened to the "Cool" Kids? Long-Term Sequelae of Early Adolescent Pseudomature Behavior." *Child Development*.
14. Kasser, T., & Ryan, R. M. (2001) "Be careful what you wish for: Optimal functioning and the relative attainment of intrinsic and extrinsic goals." *Life goals and well-being: Towards a positive psychology of human striving*. Hogrefe & Huber Publishers.

Chapter Twenty-Two: The Love Witch

1. Ehrlich, David. (November 8, 2016) "'The Love Witch' Review: Anna Biller's Technicolor Throwback Is a Spellbinding Feminist Delight." *Indie Wire*.
2. Raypole, Crystal. (February 1, 2021) "How to Recognize A Bout of Lovesickness—and What You Can Do to 'Cure' It." *Healthline*.
3. Natale, Nicol. (July 11, 2018) "7 Physical and Psychological Changes That Happen When You Fall in Love." *Business Insider*.
4. Younger, J. et al. (2010) "Viewing Pictures of a Romantic Partner Reduces Experimental Pain: Involvement of Neural Reward Systems." *PLoS ONE*.
5. (2020) "Do your eyes dilate when you are attracted to someone?" *VSP*.
6. Chapman, Gary. (1992) *The Five Love Languages: How to Express Heartfelt Commitment to Your Mate*. Northfield Publishing.
7. Engle, Gigi. (2020) *All the F*cking Mistakes: A Guide to Sex, Love, and Life*. St. Martin's Publishing Group.
8. Scarf, Maggie. (1995) *Unfinished Business: Pressure Points in the Lives of Women*. Ballantine Books.
9. Griffin-Shelley, Eric. (1997) *Sex and Love: Addiction, Treatment and Recovery*. Westport, Connecticut: Praeger.

10. Raypole, Crystal. (February 1, 2021) "How to Recognize A Bout of Lovesickness—and What You Can Do to 'Cure' It." *Healthline.*
11. Stergiopoulos, Erene. (June 29, 2016) "Real-Life 'Love Potions' Are Coming, But Are they Ethical?" *Vice.*
12. Raypole, Crystal. (February 1, 2021) "How to Recognize A Bout of Lovesickness—and What You Can Do to 'Cure' It." *Healthline.*
13. Hood, Abby Lee. (October 17, 2019) "Love Spells Aren't As Harmless As You May Think, According To Witches." *Bustle.*
14. Beverley, Robert. (2010) "Book II: Of the Natural Product and Conveniencies in Its Unimprov'd State, Before the English Went Thither". *The History and Present State of Virginia, In Four Parts.* University of North Carolina.
15. Ehrlich, David. (November 8, 2016) "'The Love Witch' Review: Anna Biller's Technicolor Throwback Is a Spellbinding Feminist Delight." *Indie Wire.*
16. (2019) "Fact or Fiction: You Can Be Prosecuted For Encouraging or Causing Someone's Suicide." *The Reeves Law Group.*
17. LePique, Annette. (June 19, 2018) "Crazy in Love: Transgressive Femininities in Anna Biller's 'The Love Witch.'" *Another Gaze.*

Chapter Twenty-Three: Penny Dreadful
1. Calia, Michael. (May 15, 2015) "'Penny Dreadful' Creator John Logan on Witches, the Occult and 'the Horror of People.'" *The Wall Street Journal.*
2. Christensen, Jen. (July 6, 2016) "What Was Behind Mary Todd Lincoln's Bizarre Behavior?" *CNN.*
3. Weiser, Kathy. (May 2021) "Native American Totem Animals & Their Meanings." *Legends of America.*
4. Bale, Christopher et al. (March 2006) "Chat-Up Lines as Male Sexual Displays." *Personality and Individual Differences.*
5. Dundes, Alan. (1998) "Bloody Mary in the Mirror: A Ritual Reflection of Pre-Pubescent Anxiety." *Western Folklore.*
6. Gregory, Andrew. (May 29, 2018) "The Best Way To Get Over a Breakup, According to Science." *Time.*
7. (2021) "Agoraphobia." *Mayo Clinic.org.*
8. Smith, Brendan L. (January 2011) "Hypnosis Today." *American Psychological Association.*
9. McFarland, Melanie. (June 30, 2016) "How Penny Dreadful's Surprise Series Finale Betrayed its Best Character." *Vox.*

INDEX

ALSO AVAILABLE

The Science of Aliens: The Real Science Behind the Gods and Monsters from Space and Time | *by Mark Brake*

The Science of the Big Bang Theory: What America's Favorite Sitcom Can Teach You about Physics, Flags, and the Idiosyncrasies of Scientists | *by Mark Brake*

The Science of Doctor Who: The Scientific Facts Behind the Time Warps and Space Travels of the Doctor | *by Mark Brake*

The Science of Fortnite: The Real Science Behind the Weapons, Gadgets, Mechanics, and More! | *by James Daley*

The Science of Harry Potter: The Spellbinding Science Behind the Magic, Gadgets, Potions, and More! | *by Mark Brake & Jon Chase*

The Science of James Bond: The Super-Villains, Tech, and Spy-Craft Behind the Film and Fiction | *by Mark Brake*

The Science of Jurassic World: The Dinosaur Facts Behind the Films | *by Mark Brake & Jon Chase*

The Science of Minecraft: The Real Science Behind the Crafting, Mining, Biomes, and More! | *by James Daley*

The Science of Monsters: The Truth about Zombies, Witches, Werewolves, Vampires, and Other Legendary Creatures | *by Meg Hafdahl & Kelly Florence*

The Science of Science Fiction: The Influence of Film and Fiction on the Science and Culture of Our Times | *by Mark Brake*

The Science of Serial Killers: The Truth Behind Ted Bundy, Lizzie Borden, Jack the Ripper, and Other Notorious Murderers of Cinematic Legend | *by Meg Hafdahl & Kelly Florence*

The Science of Star Trek: The Scientific Facts Behind the Voyages in Space and Time | *by Mark Brake*

The Science of Star Wars: The Scientific Facts Behind the Force, Space Travel, and More! | *by Mark Brake & Jon Chase*

The Science of Stephen King: The Truth Behind Pennywise, Jack Torrance, Carrie, Cujo, and More Iconic Characters from the Master of Horror | *by Meg Hafdahl & Kelly Florence*

The Science of Strong Women: The True Stories Behind Your Favorite Fictional Feminists | *by Rhiannon Lee*

The Science of Superheroes: The Secrets Behind Speed, Strength, Flight, Evolution, and More | *by Mark Brake*

The Science of Time Travel: The Secrets Behind Time Machines, Time Loops, Alternate Realities, and More! | *by Elizabeth Howell, PhD*

The Science of Women in Horror: The Special Effects, Stunts, and True Stories Behind Your Favorite Fright Films | *by Meg Hafdahl & Kelly Florence*